HARVARD CASE HISTORIES IN EXPI

JAMES BRYANT CONANT · GE



CASE 2

# *The Overthrow of the Phlogiston Theory*

## *The Chemical Revolution of 1775–1789*

EDITED BY

JAMES BRYANT CONANT

HARVARD UNIVERSITY PRESS · CAMBRIDGE · 1955

SECOND PRINTING

PRINTED IN THE UNITED STATES OF AMERICA

# *Foreword*

The Harvard Case Histories in Experimental Science have been designed primarily for students who are majoring in the humanities or the social sciences. Such students require an understanding of science that will help them to relate developments in the natural sciences to those in the other fields of human activity. To do so demands an understanding both of the methods of experimental science and of the growth of scientific research as an organized activity of society. Experience shows that a man who has been a successful investigator in any field of experimental science approaches a problem in pure or applied science, even in an area in which he is quite ignorant, with a special point of view. One may designate this point of view "understanding science"; it is independent of a knowledge of the scientific facts or techniques in the new area. Even a highly educated and intelligent citizen without research experience will almost always fail to grasp the essentials in a discussion that takes place among scientists concerned with a projected inquiry. This will be so not because of the layman's lack of scientific knowledge or his failure to comprehend the technical jargon of the scientist; it will be to a large degree because of his fundamental ignorance of what science can or cannot accomplish, and his subsequent bewilderment in the course of a discussion outlining a plan for a future investigation. He has no "feel" for what we may call "the tactics and strategy of science."

As a citizen, a businessman, a public servant, a lawyer, a teacher, or a writer, any present college student may a decade hence be called upon to evaluate the work of scientists and to consider the ways in which such work can be organized and financed. He may wish also to consider to what degree systematic investigations more or less analogous to those that have been successful in the physical sciences may be fruitful in other fields. Although the first cases are concerned almost exclusively with the physical sciences, an understanding of the variety of methods by which these sciences have advanced will provide some basis, it is hoped, from which the nonscientists can appraise proposals for research and its applications in the biological and social sciences.

## 1. UNDERSTANDING SCIENCE

In order to appreciate the methods of the experimental sciences it is necessary to master a certain amount of technical subject matter within the fields of physics and chemistry although no one can prescribe the particular topics to be studied. It is quite clear that one need not attempt the task of surveying all physics or chemistry, nor is it necessary to memorize comprehensive factual information about the scientific basis of our modern technological civilization. If one understands the nature of modern science, such factual information can when needed be readily acquired by consulting appropriate books and articles.

A direct study of the methods of modern science presents great difficulties. A visitor to a laboratory, unless he is himself a scientist, will find it almost impossible to understand the work in progress; he will comprehend neither the objectives nor the implications of the measurements and observations that the investigator is making. To be sure, he may be able to appreciate in a general way the nature of the problem that is under investigation; he may well realize that the questions raised by those who planned the experiment are of importance—either for immediate practical reasons or as part of the general advance of science. But unless he himself has had considerable experience with scientific research, he will be baffled by what he sees or hears, for almost without exception research problems today are concerned with areas of science where there is a large accumulated background of technical information. To be able to follow the new work one must have studied the advance of the science thoroughly, which usually means mastering a considerable amount of mathematics as well as physics or chemistry or biology.

Modern science has become so complicated that today methods of research cannot be studied by looking over the shoulder of the scientist at work. If one could transport a visitor, however, to a laboratory where significant results were being obtained at an early stage in the history of a particular science, the situation would be far different. For when a science is in its infancy, and a new field is opened by a great pioneer, the relevant information of the past can be summed up in a relatively brief compass. Indeed, if the methods of experimental science are being successfully applied for the first time to a problem of importance, the scientist's knowledge would not be much greater than that of his inquiring guest. Briefly, and in simple language, he could explain the new experiment. Then as from day to day results were obtained and further experiments planned the visitor would see unfolding before him a new field. However, such periods in the history of science are relatively few, and it would be a fortunate visitor indeed who could spend some time merely as a witness of the events. For he would have the pleasure of knowing not only that he had been present at one of the critical battles in the forward march of science, but that he had had a unique opportunity of learning at first hand about the methods of science.

### 2. THE USE OF CASE HISTORIES

The purpose of the case histories presented in this series is to assist the student in recapturing the experience of those who once participated in exciting events in scientific history. The study of a case may be to some degree the equivalent of the magical operation suggested in the preceding paragraph, namely, that of transporting an uninformed layman to the scene of a revolutionary advance in science. To be sure, usually more than one investigator or laboratory is involved and the period of time is at least several years; for, however significant a single experiment may seem in retrospect, no important step forward in experimental science rests solely on the record of any single investigator's observations. Rather, the interplay of ideas of several men, arguments about experiments and their interpretation between workers in different laboratories, often mark the decisive turn in scientific thinking. However, by the intensive study of the actual work of some of the great investigators of the past (and in less detail, the work of their contemporaries), the methods then employed stand out clearly.

Some of the cases in this series present the work of men who lived over three hundred years ago; others are drawn from the eighteenth and nineteenth centuries; a few may involve twentieth-century discoveries. But irrespective of their dates, the examples presented illustrate the methods of modern science. Familiarity with those methods will increase the ease of understanding of the work of scientists today; as one consequence, popular accounts of scientific progress can be read with more appreciation. If one is face to face with problems of financing or organizing research, a critical but sympathetic attitude toward experimental investigation is a prerequisite for an intelligent examination. The occasion may arise in connection with private industry, the expenditure of public money, or a philanthropic enterprise related to medicine and public health. Almost all that a trained scientist has to go on when he passes judgment on the prospects of a new venture far removed from his own specialty is his knowledge of the methods by which science has progressed in his own experience. To a considerable degree a nonscientist may come to have a similar basis for his opinions. An intimate acquaintance with a relatively few historic cases should assist him in finding his way through the complexities of modern investigation as he listens to those who tell him what is being proposed.

### 3. CONCERNING THE METHODS OF SCIENCE

Science we shall define as a series of concepts or conceptual schemes arising out of experiment and observation and leading to new experiments and new observations. From the experimental work and careful observations of nature come the scientific facts that are tied together by the concepts and conceptual schemes of modern science. The rather sudden burst of interest in the new "experimental philosophy" in the seventeenth century is historically related to a new interest on the part of thoughtful men in practical matters ranging from medicine through mining to the ballistics of cannon balls. But while the problems were often suggested by the interest of learned men in the practical arts, the development of science involved something more than the type of experimentation by which the practical arts had been developing for centuries. The method of reasoning employed in mathematics (for example, in geometry), commonly called deductive reasoning, had to be combined with the methods of experimentation that came from the practical arts before science could progress rapidly. The interaction of these two streams of human activity was largely responsible for the development of physics and chemistry; the focus of attention was shifted from an immediate task of improving a machine or a process to a curiosity about the phenomena in question. New ideas or concepts began to be as important as new inventions, and their interweaving began to produce conceptual fabrics whose various threads gained support from one another as well as from the direct results of experiment and observation.

*Concepts and Conceptual Schemes.* We have defined science as a series of concepts or conceptual schemes (theories) arising out of experiment or observation and leading to new experiments and observations. The cases presented in this series illustrate some of the many ways in which the new ideas (concepts) have arisen and the consequences they have entailed. In more than one case we shall see a new conceptual scheme (theory) replacing an older one, or two rival schemes in conflict. The transition to a new theory is seldom easy; old ideas are apt to be tenacious. Looking back at the history of any branch of science we can see that a new conceptual scheme (theory) comes to be accepted because it is at least as satisfactory as the older one in relating all the known facts in a simple way and because it proves to be more fruitful of new experiments. The latter outcome is clearly a matter for the future, so that there can be no way to determine in advance whether a new concept or conceptual scheme will prove acceptable in this double sense; nor can one foretell how soon the next major advance will occur. A study of these cases makes it clear that there is no such thing as *the* scientific method, that there is no single type of conceptual scheme and no set of rules specifying how the next advance will be made through the jungle of facts that are presented by the practical arts on the one hand and by the experimentation and observation of scientists on the other.

*Speculative Ideas and Broad Working Hypotheses.* The relation between a speculative idea and a broad working hypothesis that becomes a wide conceptual scheme is of interest. A good example is furnished by the history of the atomic theory. The notion that there are fundamental units — ultimate particles — of which matter is composed goes back to ancient times. But expressed merely in general terms this is a speculative idea until it forms the basis of a broad working hypothesis from which consequences can be deduced that are capable of experimental test. This particular speculative idea became in Dalton's hands a broad working hypothesis and came to be a new conceptual scheme only after he had shown, about 1800, how fruitful it was in connection with the quantitative chemical experimentation that had been initiated by the chemical revolution.

Dalton's atomic theory is an instance where we can trace the origin of a conceptual scheme; in other cases we are uncertain how the idea came to the proponent's mind. Thus Torricelli seems to have been the first to suggest that we live in a "sea of air" that exerts a pressure analogous in many ways to the pressure of water on submerged objects; yet we have insufficient knowledge to say what led him to this idea. His famous experiment with a column of mercury, in which he made the first barometer, may have been the point of departure or it may have been a case of testing a deduction from a broad working hypothesis, arrived at by means that history does not record. At all events, from that time on a number of deduced consequences led to experiments nearly all of which were consistent with the ideas of Torricelli, and his hypothesis quickly attained the status of a new conceptual scheme.

The cases presented in this series emphasize the prime importance of a broad working hypothesis that eventually becomes a new conceptual scheme. They illustrate the variety of mental processes by which the pioneers of science developed their new ideas. But it must be emphasized that great working hypotheses have in the past often originated in the minds of these scientific pioneers as a result of mental processes that can best be described by such words as "inspired guess," "intuitive hunch," or "brilliant flash of imagination." The origins of the working hypotheses are to be found almost without exception in previous speculative ideas or in the previously known observations or experimental results. Only rarely, however, do these broad working hypotheses seem to have been the product of a careful examination of all the facts and a logical analysis of various ways of formulating a new principle.

*The Testing of Deductions from Conceptual Schemes.* When first put forward, a new conceptual scheme may be no more than a working hypothesis on a grand scale. From it we can deduce, however, many *consequences*, each of which can be tested by experiment. If these tests confirm the deductions in a number of instances, the accumulating evidence tends to confirm the broad working hypothesis and the hypothesis may soon come to be accepted as a new conceptual scheme; its subsequent life may be short or long, for from it new consequences are constantly being deduced that can be verified or not by experimental test.

While the modes of formulating broad working hypotheses are so varied as to defy generalization, the procedures by which these hypotheses are tested conform to a fairly consistent pattern. One is tempted, therefore, to represent the process of testing a broad conceptual scheme by a diagram. Unfortunately, however, the interrelations are so complex as to be more confusing than helpful. One cannot grasp an entire case at a glance; to get the real point of the story it must be studied in a step-by-step fashion at a reasonable pace. These Case Histories in Experimental Science provide the material for just such study.

While our purpose would be defeated in the distortion necessary to compress a case history into a single diagram, we can perhaps suggest by a particular example, drawn from *part* of one case, the type of process involved in the experimental testing of broad conceptual schemes. Thus, at one stage in the development of pneumatics, Pascal deduced from Torricelli's idea of a sea of air the consequence that the air pressure should be less on a mountain top than at the base. This deduction could be expressed in some such words as these: if the Torricellian experiment is performed at such and such a place the height of the mercury column will be greater than on the same day at a nearby spot high up on a mountain. This prediction is susceptible of a test which yields a yes or no answer.

As the cases illustrate, however, there are often many slips in the argument at this point; unconscious assumptions are often involved that turn out to have been unwarranted. In general, the complete pattern of deduction linking the broad working hypotheses to the experimental tests proves far from simple. Thus, in the example just cited, the tacit assumption is clearly being made that the laws of liquid pressure and the density of mercury (and probably several other such things as well) are not themselves appreciably different at different altitudes; these assumptions find their ultimate justification (for the purposes of *this* experiment, at least) in other broad conceptual schemes such as make up the laws of hydrostatics. It is the multiplicity, the complexity, and indeed often the vagueness of this sort of interconnection that is ignored in many oversimplified discussions of "*the* scientific method."

*Experimentation.* The experiments performed to test deductions from broad working hypotheses involve in the last analysis a type of activity with which everyone is familiar. Examples of experimentation in everyday life come readily to mind. Whenever one has a choice of alternative ways of trying to solve a problem, decides on one, and then says in essence "let's try it and see," he is getting ready to perform an experiment. Framing various possibilities as to what is wrong with an automobile engine when the car is stalled is usually a preliminary to carrying out an actual trial, an experiment. Similarly, if a man is trying to locate a broken wire, a blown-out fuse, or a bad tube in a radio set, he usually turns over in his mind a number of tests that could be made and then tries first one and then another. He is guided in the formulation of his trials (experiments) by his knowledge of the automobile or the radio set in question and his general information about the behavior of these machines. When he finally carries out the experiment he is, however, testing the validity of a very *limited working hypothesis.* This limited working hypothesis can generally be stated in some such form as "if I do so and so, then such and such will happen." A simple and perhaps trivial example of the testing of a limited working hypothesis is when a person trying to unlock a door puts in the lock a hitherto untried key and says to himself, "if I turn the key, then the lock will spring." The trial either validates or negates the limited working hypothesis; the same is true of the final step in most experiments in science.

The difference between experimentation in everyday life and experimentation in science lies primarily in the fact that in the one case common-sense assumptions and practical experience determine the nature of the experiment; in the other, a series of connecting links usually relates some deductions of a broad working hypothesis with the final limited working hypothesis involved in the specific experiment. This will be more evident as one studies the cases. The steps by which practical men throughout the ages have improved the practical arts have also involved experimentation. Here, as in everyday life, however, the experiment is in almost every case planned in terms of common-sense assumptions and practical experience, not in terms of broad generalizations or conceptual schemes. Furthermore, as was pointed out in the opening paragraph of this section, experimentation in the practical arts always has a practical objective. Experimental science was the consequence of the combination of the ancient art of experimentation with the type of thinking that employs general ideas and deductive reasoning. Experimentation by itself does not produce science, although it is an essential element in advance in many fields of science.

The progress of science in many areas has depended on the overcoming of experimental difficulties, in part by the invention of new instruments and the discovery of new materials. A study of the types of difficulties that have arisen in the past and how they have interfered with the testing of new concepts and conceptual schemes throws light on what is going on in the laboratories of physics and chemistry today. Furthermore, in formulating problems according to certain conceptual schemes, and particularly in planning the experimental tests, it became necessary as science advanced to make precise and accurate many vague common-sense ideas, notably those connected with measurement. Old ideas were clarified or new ones introduced. These are the new concepts that are often quite as important as the broad conceptual schemes; for example, pressure, acceleration, mass, temperature, atom, are words that have now precise scientific meanings. At the time the new ideas were advancd that are considered in the case histories from the seventeenth, eighteenth, and nineteenth centuries, they were as novel as are in our own times the ideas of neutrinos and mesons in the field of nuclear physics.

A close attention to the difficulties that have beset investigators in getting clean-cut answers to apparently simple questions will suggest some of the complexities of research today. And the broader hypotheses must remain speculative ideas until one can by the processes indicated succeed in relating them to experiments.

### 4. THE ADVANCE OF SCIENCE AND PROGRESS IN THE PRACTICAL ARTS

The relation between advances in the practical arts (such as mining, agriculture, manufacturing) and progress in science is more complicated than at first sight appears. To appreciate the present situation, it is helpful to have some knowledge of the history of these two branches of human endeavor and of their interaction. For many generations after the new experimental philosophy was put forward by a few amateurs in the seventeenth century, the direct impact of science on the practical arts was almost negligible. This was the consequence not of any lack of interest in practical affairs on the part of the early scientists but of the fragmentary nature of the knowledge they acquired and the inadequacy of the conceptual schemes they elaborated. The advances in the practical arts of mining, metal making, agriculture, food processing, and medicine in the eighteenth and early nineteenth centuries were made without any very direct benefit of science. (The art of navigation is an exception to this statement.)

The ancient method of experiment, the use of the "let's-try-it-and-see" type of reasoning in solving immediate practical problems, has led over the centuries to amazing results. This process of advance is often called *pure empiricism*, there being no wide generalization to guide the inventor. Even today in the mid-twentieth century when science has penetrated so deeply into industry, there is still a great deal of empiricism in a practical art such as steel making. Even when no practical aim is involved, the practical methods of a science may contain many recipes and rules that have no relation to theory; in other words, there is some empiricism in almost all scientific procedures even now. It is often convenient to speak of the "degree of empiricism" in a given industrial process or branch of science. Where wide generalizations and theory enable one to calculate in advance the results of an experiment or to design a machine (a microscope or a telescope, for example), we may say that the degree of empiricism is low. On the other hand, if the conditions for satisfactory operation of a machine, a process, or an instrument are based only on cut-and-try experimentation (as is true in some branches of chemistry), we may say that the degree of empiricism in that branch of science is still high.

### 5. PURE AND APPLIED SCIENCE

Applied research may be defined as research directed to the end of reducing the degree of empiricism in a practical art. Activities concerned solely with developing and improving our conceptual schemes and filling in gaps in our systematic knowledge may be defined as pure science. Modern science is a fabric the texture of which is composed of many interrelated concepts and conceptual schemes. Today the validity of each conceptual scheme must be tested in the light of our total knowledge. The cases presented in this series fall in the category of pure science; they are nevertheless concerned for the most part with episodes in the history of science that eventually had great influence on applied science. The conceptual schemes elaborated by scientists have been fundamental to further discoveries and to the formulation of the ideas that are the fabric of modern science. With the aid of the concepts and theories developed over the past three hundred years, and drawing on the systematic knowledge accumulated during that period, the applied scientist today can improve the industrial arts and invent new processes, products, and machines far more readily than could his predecessors of the seventeenth and eighteenth centuries. Science today is closely connected with technology because theory and the practical arts have become closely interwoven. The degree of empiricism has been lowered on almost every hand and is being still further reduced by the activity of investigators in laboratories of both pure and applied science. The applied scientist or engineer can now engage in what was once an activity of those inventors who proceeded largely empirically. The application of science to new industrial processes is often designated "development work"; applied research and development today go hand in hand.

The analysis of experimental science presented in this foreword and a study of the cases outlined in this series will, it is hoped, lead the layman to a better understanding both of modern science and of modern technology. No college course can be a substitute for experience; it can only lay a foundation. With an introduction to an understanding of experimental science, the inquiring citizen ought to be in a position to continue his scientific education throughout his life. In the belief that a detailed knowledge of a few epoch-making advances in science will provide a key to a better comprehension of the modern world the Harvard Case Histories in Experimental Science are presented to the public.

In the preparation of this Foreword and of the first two Cases of the series, the editor has benefited greatly from the suggestions put forward during discussions of the presentation of this and similar material in a General Education course in Harvard College. These discussions have continued almost weekly over several years and it would be impossible to make acknowledgment of the specific contributions of the various members of the group. The editor would like, however, to express his gratitude for the assistance of Professor Duane Roller of Wabash College, Professor Gwilym E. Owen of Antioch College, Associate Professor Fletcher G. Watson of Harvard, Assistant Professor Leonard K. Nash of Harvard, Dr. Charles L. Clark, Dr. Thomas S. Kuhn, and Dr. Eugene P. Gross.

To Mr. Duane H. D. Roller the editor is particularly indebted for his careful handling of the proofs and his editorial assistance.

Mr. J. D. Elder of the Harvard University Press has been helpful in many ways.

To all of these gentlemen the author wishes to express his thanks.

JAMES B. CONANT

*Cambridge, Massachusetts*
*March 31, 1950*

CASE **2**

# *The Overthrow of the Phlogiston Theory*

## INTRODUCTION

The overthrow of the phlogiston theory involved the development of a superior conceptual scheme. There can be no doubt that the author of the new ideas was the French chemist Lavoisier, often referred to as the "father of modern chemistry." The chemical revolution did not occur overnight, however, and Lavoisier did not accomplish his results singlehanded. By studying the evolution of a new conceptual scheme we can see the difficulties with which pioneers in science almost always have to contend, and the false steps that usually accompany even the most successful forward marches. We likewise see how, by the end of the eighteenth century, science had become an organized activity with its own methods of making public new results. As compared with the first half of the seventeenth century, the interchange between leading investigators was prompt and effective. As compared with our own time, or even with the middle of the nineteenth century, however, the publication of results was sporadic and incomplete. For example, although the English experimenters Priestley and Cavendish could read of Lavoisier's work soon after it was completed (and vice versa), the fact that the Swedish chemist Scheele had prepared pure oxygen gas and studied its properties remained unknown to what was then the "scientific world" for several years. Scheele's experiments were completed before 1773. His characterization of a new gas which he called "fire air" was quite unambiguous. The manuscript of his book on *Air and Fire* describing these discoveries and interpreting them was not delivered to the printer before 1775 and not published until 1777. This delay prevented his discovery of oxygen, which antedated Priestley's, from being the effective discovery that influenced the work of other men.

The chemical revolution may be conveniently thought of as occurring in the following stages:

(1) Lavoisier's experiments with sulfur and phosphorus in 1772 in which he noted a "prodigious quantity of air was fixed."

*(2) The effective discovery of oxygen, the unraveling of its relation to the calcination of metals in air, and its role in the respiration of animals, 1774–1778.

(3) The determination of the composition of water, 1782–1783.

(4) The extension and precise formulation of the new conceptual scheme in terms of a new chemical nomenclature in Lavoisier's *Traité Élémentaire de Chimie,* 1789.

In this Case we shall center our attention on the second of these stages. In many ways this is the most interesting part of the entire story, for one can trace here the stumbling way in which two great scientists—Joseph Priestley (1733–1804) the English nonconformist clergyman, and Antoine Lavoisier (1743–1794) of Paris—proceeded through the jungle of new experimental facts. One can see how the publications of the two men influenced the course of scientific history. In the case of Lavoisier we can follow his line of reasoning and see his errors particularly clearly because of the way in which his most important single paper was given to the world. On the threshold of success, he is diverted by a faulty interpretation of an experiment and recovers his sense of direction only after some months of labor and with the assistance of the English chemist. We are here permitted to follow the course of a revolutionary scientific discovery in much the same fashion as a slow-motion picture of a critical moment enables us to understand what happened in a football game.

The chemical transformations involved in this case are few, and in the light of modern chemistry they are relatively simple. It will assist the student of this example of "science in the making" to master first a few chemical facts as they are expressed in terms of the conceptual scheme that has been the gospel of chemists since Lavoisier's final triumph.

Air is primarily a mixture of two gases, oxygen and nitrogen. Combustion and respiration involve chemical reactions between carbon compounds and oxygen; the products of these reactions are water and carbon dioxide, except in the case of charcoal, when carbon dioxide alone is formed. When a metal is heated in air, it forms an oxide by combining with the oxygen; the product was known to the early chemists as a "calx" and the process as "calcination." When many oxides of metals are heated with charcoal, the oxygen combines with the charcoal forming carbon dioxide ("fixed air" to the chemists of the eighteenth century) and the metal. Mercury oxide, a red powder, also known as red precipitate or *mercurius calcinatus per se,* has the unusual property of being converted into the metal mercury and oxygen when heated quite hot *without charcoal*. Therefore, one can start with mercury metal and obtain the oxide by heating the metal in air, and then by heating the oxide still hotter, regenerate the mercury. If one arranges to collect the gas evolved in this last process, one will have oxygen gas at one's disposal. The three reactions with which Lavoisier and Priestley were concerned were:

**Calcination**

| $2Hg$ | $+$ | $O_2$ | heated $\longrightarrow$ | $2HgO$ |
|---|---|---|---|---|
| mercury metal | plus | oxygen gas | yields | oxide of mercury (red powder) |

**Decomposition of oxide**

| $2HgO$ | heated very hot $\longrightarrow$ | $2Hg$ | $+$ | $O_2$ |
|---|---|---|---|---|
| oxide of mercury | yields | mercury metal | plus | oxygen gas |

**Reduction with addition of charcoal** (also called reduction with phlogiston!)

| $2HgO$ | $+$ | $C$ | heated $\longrightarrow$ | $2Hg$ | $+$ | $CO_2$ |
|---|---|---|---|---|---|---|
| oxide of mercury | plus | charcoal (carbon) | yields | mercury metal | plus | carbon dioxide or "fixed air" |

*The Phlogiston Theory*. A few words about the phlogiston theory are necessary by way of introduction. Joseph Priestley never gave up his belief in the theory. In his last book, published in 1796, he wrote, "There have been few, if any, revolutions in science so great, so sudden, and so general, as the prevalence of what is now usually termed the *new system of chemistry*, or that of the *Anti*phlogistons, over the doctrine of Stahl, which was at one time thought to have been the greatest discovery that had ever been made in the science. I remember hearing Mr. Peter Woulfe,[1] whose knowledge of chemistry will not be questioned, say, that there had hardly been any thing that deserved to be called a *discovery* subsequent to it. Though there had been some who occasionally expressed doubts of the existence of such a principle as that of

[1] Chemist and mineralogist, born 1727, died 1803, resident of London and Paris, member of the Royal Society from 1767.

*phlogiston*, nothing had been advanced that could have laid the foundation *of another system* before the labors of Mr. Lavoisier and his friends, from whom this new system is often called that of the *French*."

The significance of this quotation lies in the admission by Priestley, the die-hard upholder of the phlogiston theory, that Lavoisier and his friends had brought about a chemical revolution. Although still raising doubts about the new views and retaining the vocabulary of the phlogiston theory, Priestley in this last publication seems to be admitting that he is fighting a rear-guard action. His statement as to the former high position held by Stahl's doctrine of phlogiston is confirmed by the writings of many eighteenth-century scientists. The phlogiston theory was a broad conceptual scheme into which could be fitted most of the chemical phenomena of the mid-eighteenth century. The lineage of this theory can be traced back to the alchemists. (It should be noted that a metal, according to this theory, is more complex than is the corresponding oxide.) In particular, it accounted for one of the simplest chemical processes then employed for practical ends, namely, the preparation of metals from their ores. The transformation of an earthy substance into a metal in the smelting process appeared to be much the same whether the metal was iron, or tin, or copper. What could be more plausible than to assume that in each instance the ore, when heated with charcoal, took up a "metallizing principle" which conferred upon the earth the properties of a metal? If one called this hypothetical substance phlogiston, an "explanation" for metallurgy was at hand:

| Metallic ore (an oxide) | + | Phlogiston from charcoal | ⟶ | Metal |
|---|---|---|---|---|

The fact that charcoal would burn by itself when heated indicated to the founders of the phlogiston theory that the phlogiston escaped in the process and became combined with the air. In general, substances that burned in air were said to be rich in phlogiston; the fact that combustion soon ceased in an enclosed space was taken as clear-cut evidence that air had the capacity to absorb only a definite amount of phlogiston. When air had become completely phlogisticated it would no longer serve to support combustion of any material, nor would a metal heated in it yield a calx; nor could phlogisticated air support life, for the role of air in respiration was to remove the phlogiston from the body. Everything fitted together very well.

The phlogiston theory was almost universally accepted at the time of the American Revolution and was the basis of the chemistry then taught to college students as part of their instruction in natural philosophy. The lecture notes of Professor Samuel Williams, the Hollis Professor of Mathematics and Natural Philosophy, 1780–1788, at Harvard illustrate the convincing way in which the phlogiston theory could be presented to a class. Presenting the subject of Pneumatics, Professor Williams listed six different airs, and about common or atmospherical air had this to say:

"Among the various kinds of permanently elastic Fluids we may begin with *Common or Atmospherical Air*. Atmospherical Air is that which we breathe, with which we are constantly surrounded, and which is common to every country and place. And with regard to this kind of Air several observations may be made. Thus: Common or atmospherical Air may combine, or be charged with many substances.—Thus water is held in solution by Air, and Rain, Snow, and Hail appear to be precipitations from it. . . . But what is worthy of a particular observation, common or atmospherical Air is generally charged with a large quantity of *Fire* or *Phlogiston*. The Chemists sometimes speak of Fire and Phlogiston as being the same thing, or signifying the same element. But we are not absolutely certain that this is the case. By Phlogiston we mean no more than *the principle* of Inflammability; or that by which bodies become combustible or capable of burning.—And that there is such a principle or element as Phlogiston, and that common Air may be charged with large quantities of it may be easily represented.

"*Example:* Take some combustible substance and let it be inflamed or set on fire: In this state inclose it in a vessel containing a small quantity of atmospherical air. *Effect.* The combustion will continue but a small time and then cease. Part of the combustible substance is reduced to ashes and the other part remains entire. And the Air appears to be changed and altered . . . Here then we have a representation of what the chemists call Phlogiston and of the Air's being loaded with it. In the confined air the combustible matter continues burning until the air becomes loaded with something that prevents any further combustion. And being confined by the closeness of the vessel whatever the matter be with which the air is loaded it is confined within the vessel and cannot escape. . . .

"It seems therefore from this Experiment that Phlogiston must be a real Substance, and that the Air is loaded or saturated with it. For what can the enclosing the combustible matter in the Phial do but to prevent the escape or dispersion of some real substance? And is it not evident that so long as the air can receive this substance from the combustible matter so long the body will continue burning; and that as soon as the Air is saturated and can receive no more of the Phlogiston, the combustion must cease for no more Phlogiston can escape or be thrown out from the burning body. And therefore when fresh air is admitted to receive Phlogiston, the combustion will again take place.— And hence are derived the phrases of *phlogisticated* and *dephlogisticated* air. By phlogisticated air is intended air which is charged or loaded with Phlogiston, and by dephlogisticated air is meant Air which is free from Phlogiston; or which does not contain this principle or element of inflammability."

*Qualitatively,* the phlogiston theory was a satisfactory framework to accommodate the chemical phenomena known in the 1770's. Even some quantitative changes could be accounted for as explained by Professor Williams later in the same lecture:

"We generally estimate the purity of the air by its fitness for the purposes of combustion, respiration, and animal life. And the more free it is from Phlogiston, the principle of inflammability, or any noxious effluvia or mixture, the more fit it is found for this purpose . . . *Experiment.* Take a receiver which holds one gallon and put a lighted candle into it, and let the air be confined.—*Effect:* The candle will burn dimmer and dimmer until the light goes out. Observe the time, and this will be found to be in about one minute after it was put in.— *Remarks:* It appears from this experiment that about one gallon of common air is requisite to support so small a fire or flame as that of this candle for the space of one minute. In this time the quantity of air is diminished by about 1/15 or 1/16 of its bulk."

The diminution in bulk of the air, Professor Williams explains, is a consequence of the phlogistication of the air. For he says, "The greatest diminution of air by Phlogiston is about one-fourth of its first quantity and air which is diminished to its utmost by any one process cannot be any further diminished by another."

*Lavoisier's Broad Working Hypothesis.* The increase in weight of metals on calcination was, however, a quantitative observation that presented great difficulties to those who thought in terms of phlogiston. After the discovery of the compound nature of water, an explanation was contrived, but it had only a short life, for the phlogiston theory was then going to pieces rapidly.

Lavoisier's new system of chemistry seems to have started with his pondering on the very large increase in weight when phosphorus was burned in air. In a sealed note deposited with the Secretary of the French Academy on November 1, 1772, Lavoisier wrote:

"About eight days ago I discovered that sulfur in burning, far from losing weight, on the contrary, gains it; it is the same with phosphorus; this increase of weight arises from a prodigious quantity of air that is fixed during combustion and combines with the vapors.

"This discovery, which I have established by experiments, that I regard as decisive, has led me to think that what is observed in the combustion of sulfur and phosphorus may well take place in the case of all substances that gain in weight by combustion and calcination; and I am persuaded that the increase in weight of metallic calxes is due to the same cause."

We seem to see here the flash of genius that puts forward a bold working hypothesis on a grand scale without much evidence to support it. Yet there is no doubt, as Lavoisier always claimed, that the essential idea in his theory was contained in this note; *something* was taken up from the atmosphere in combustion and calcination. This was exactly opposite, be it noted, to the phlogiston doctrine. But what was the something? Lavoisier experimented with gases in a search for the answer without much success until the winter of 1774–1775. His book recounting his experiments (largely a repetition of the work of Joseph Black [2] and Priestley), published in 1774, gives the details of his experiment with phosphorus. (He never reported further on his experiments with sulfur, which were in part, at least, erroneous.) He convincingly demonstrates that the white solid produced when phosphorus burns in air confined over mercury weighs about twice as much as the phosphorus that produced it. But as to the nature of the "prodigious quantity of air" that is fixed in the process he remains uncertain. It was the experiments with red oxide of mercury that first put him on the right track. Perhaps it is not too much to say that the unique properties of this oxide made possible a relatively rapid advance in the chemistry of combustion. For no other oxide among all those known even today has the properties of (*a*) decomposing into the metal and oxygen when heated alone below "red heat"; (*b*) being readily prepared free from carbonate; (*c*) being formed when the metal is heated in air at atmospheric pressure.

For a successful investigation a chemist should have pure chemical compounds; mixtures of compounds are very baffling. Furthermore, he must study, if possible, relatively simple reactions that yield only a few products. If by changing the conditions the reaction can be reversed (i.e., be made to run in the other direction) so much the better, for he can then proceed by synthesis and analysis. For example, in the

[2] Joseph Black (1728–1799), Professor of Anatomy and Chemistry at Glasgow, later Professor of Chemistry at Edinburgh. His work on latent and specific heats (see Case 3 of this series) was of first importance in the development of eighteenth-century physics. His investigation of the process of preparing lime from limestone was a model for later chemical investigations; from him Lavoisier learned the significance of the quantitative study of a simple chemical reaction (limestone → lime + carbon dioxide) that could be reversed.

case of the red oxide of mercury, he can start with mercury and oxygen and prepare the oxide, or conversely, he can start with the red oxide and prepare mercury and oxygen.

### 1. THE OPENING PHASE OF THE CHEMICAL REVOLUTION

We now come to the crucial events in the chemical revolution (the overthrow of the phlogiston theory), namely, the effective discovery of oxygen. The following chronology of events may prove a useful guide.

February 1774. The French chemist, Bayen, calls attention to the fact that red oxide of mercury yields mercury when heated alone (no charcoal needed as with other calxes). He erroneously identifies the gas evolved as "fixed air" (carbon dioxide)!

August 1774. Priestley prepares oxygen by heating red oxide of mercury, but mistakes the new gas for "laughing gas."

October 1774. Priestley tells Lavoisier of his experiment.

March 1775. Priestley makes the effective discovery of oxygen.

Easter 1775. Lavoisier communicates his memoir "On the Nature of the Principle which Combines with Metals during Calcination and Increases their Weight" to the French Academy.

November 1775. Priestley corrects, in print, Lavoisier's faulty interpretation of the experiment.

May 1777. Lavoisier communicates to the French Academy his memoir on respiration of animals; his ideas about oxygen are now clear.

August 1778. Lavoisier publishes his revised "Easter Memoir."

In order to read intelligently the papers by Lavoisier and Priestley dealing with the discovery of oxygen, a few facts about the oxides of nitrogen must be at hand. These facts are related to Priestley's "nitrous air test" for the purity of common air. Priestley had some years before prepared an oxide of nitrogen that plays an important part in this story; it is nitric oxide (NO), called by Priestley and his contemporaries "nitrous air." Priestley knew that this colorless gas, which is insoluble in water, when mixed with air produced a red gas ("red fumes") that was soluble in water. He found that air in which a candle had been burned until the flame went out would not produce soluble red fumes with "nitrous air." The reason for this, we now know, is that the reaction is between the "nitrous air" and the oxygen:

| $2NO$ | $+$ | $O_2$ | $\longrightarrow$ | $2NO_2$ |
|---|---|---|---|---|
| "nitrous air" | plus | oxygen | yields at once | "red fumes" that dissolve in water |

Since the product of this almost instantaneous reaction is soluble in water, it is evident that if the two gases are mixed over water there will be a diminution in volume. Indeed, if only the two pure gases were at hand and the volumes were chosen correctly, there would be no residual gas left after the reaction; the diminution would be complete. Since air is only about one-fifth oxygen, there will always be a large amount of residual gas when "nitrous air" and common air are mixed; the nitrogen does not react and is only very slightly soluble in water.

Priestley conceived the idea of using this reaction to test the "goodness" of common air. He found by repeated trials that with ordinary air (the best he knew) the maximum contraction occurred when one volume of "nitrous air" was mixed with two volumes of common air over water. Under these conditions the residue, measured a few minutes later, was only about 1.8 volumes. That is, the diminution in volume was as great as the whole volume of added "nitrous air" and about 20 percent more! With air that was thoroughly "bad" the final volume would be equal to the volume of the "nitrous air" plus the common air (three volumes). Intermediate values represented to Priestley intermediate grades of "goodness." This test was well known among chemical investigators since Priestley had published his results in 1772. It was therefore natural that not only Priestley himself but Lavoisier should apply the "nitrous air test" to any new "air" that would support combustion.

Both Priestley and Lavoisier were at first misled by the application of this test to pure oxygen. Priestley corrected his own error, but Lavoisier interpreted the results correctly only after Priestley had publicly shown where the trouble lay. From the preceding paragraphs it will be evident to the reader that "nitrous air" and pure oxygen react, and when the reaction is carried out over water a large contraction in volume occurs. If Priestley's standard procedure for testing the goodness of air is followed, one volume of "nitrous air" will be added to two volumes of oxygen. Under these conditions all the "nitrous air" is used up but a large amount of oxygen is left over. The actual diminution in volume will be deceptively similar to that found when common air is at hand, but the residual gas instead of being nitrogen is oxygen. There is a small but significant quantitative difference. With common air the resulting gas—the residue—occupies only 1.8 volumes if 2.0 volumes are initially employed; with pure oxygen the final volume is nearer 1.6 volumes. The fact that these numbers are as close as they are is more or less an accident since Priestley chose his standard operating procedure purely empirically (by trial and error); with larger amounts of "nitrous air," the difference between common air and oxygen would be more apparent, as will be evident in a moment. At all events, both Priestley and Lavoisier overlooked the clue offered to them by the somewhat larger diminution in volume of the new gas when subjected to Priestley's test for its "goodness."

The striking contrast between the behavior of common air and oxygen when treated with "nitrous air" over water lies in the properties of the residual gas. (This can be strikingly shown as a lecture-table demonstration, using graduated cylinders as containers in a pneumatic trough.) In the first case, nitrogen mixed with a little "nitrous air" is left; in the other, oxygen. Therefore, the residue in the one instance will support neither combustion nor animal life, nor will it react further with "nitrous air." In the other, the residue has all the properties of the original sample. Any one of the simple tests will at once make this striking difference apparent; a lighted candle, a live mouse, or the addition of "nitrous air" will convince anyone that the two samples of residual gas were totally different. But it never occurred to Lavoisier to try any of these experiments; nor did it occur to Priestley either. Only by the "accident" he describes (see Sec. 4) did Priestley come to examine what was left over when "nitrous air" had diminished his new air from red oxide of mercury! The fact that both investigators took the wrong turn in the road at a critical point in a study of the first importance illustrates how much more complicated is the advance of science than "collecting the facts, classifying the facts, formulating laws, and elaborating from the laws adequate theories."

In comparing the work of Priestley and Lavoisier it is well to bear in mind the differences in their scientific experience up to the year 1774. Priestley was then 41 years old and had started his study of "airs" (gases) some seven years before with an investigation of the fixed air generated in a brewery near his house in Leeds (where he was then the minister of a "dissenting" chapel). In the meantime, he had become the leading pneumatic chemist of the day, having improved the equipment for handling gases and discovered a number of new gases. Lavoisier, on the other hand, was not only ten years younger but a newcomer to the field of experimental chemistry who had only within the past two years repeated many of the early experiments on gases in order to learn the necessary techniques. Yet both the experienced older man and the brash young, inexperienced experimenter made essentially the same mistake and almost at the same moment. In both instances unconscious assumptions invalidated their chains of reasoning.

---

A few words are now in order about the two versions of Lavoisier's first communication of his experiments on oxygen. Lavoisier read his paper "On the Nature of the Principle which Combines with Metals during Calcination and Increases their Weight" for the first time to the French Academy of Science on Easter, 1775. (It is sometimes referred to as the Easter Memoir of 1775.) It was immediately published, presumably in the form in which he read it, in the May 1775 number of the *Journal de Physique* (page 429). (This magazine was edited by Rozier and carried the full title *Observations sur La Physique, sur L'Histoire Naturelle et sur Les Arts et Métiers.*) The official publication of the Academy of Sciences, the *Mémoires de l'Académie des Sciences,* was at that time months or years late in appearing; this fact gave the authors ample opportunity to revise their communications before their final and authoritative printing. Lavoisier's Easter Memoir of 1775 was again read to the Academy on August 8, 1778 and reëdited at that time, incorporating a few very important changes. A comparison between the text in Rozier's *Journal* of May 1775 and the official paper published after 1778 affords the rare opportunity previously mentioned of viewing a great discovery in the making.

In the spring of 1775 Lavoisier just missed seeing the full significance of the experiments with the red precipitate of mercury (oxide of mercury). He appears to have thought that he had common air at hand rather than a new gas; it was the experiment with Priestley's method of testing the "goodness" of an air by the use of "nitrous air" that was primarily responsible for his erroneous conclusion. For the sixth point in the first version of his paper (May 1775) states that the gas evolved from red oxide of mercury "was diminished like common air by an addition of a third of nitrous air." Priestley, on reading this sometime in the summer of 1775, hastened to record in a book of his, printed a few months later, that this was wrong (see Sec. 3). We cannot be certain that these words of his rival put Lavoisier back on the right track, but we find from Lavoisier's laboratory notebooks that in the following February (1776) he had prepared an "air" from a sample of red oxide of mercury and found it to be considerably "better" than common air *by the test with "nitrous air."* In May 1777, Lavoisier read to the Academy a paper on the respiration of animals in which he makes clear that air is a mixture of two gases, one "highly respirable," the other unable to support combustion or respiration. By 1778 there was no doubt in anyone's mind that a new gas, *not* common air, was produced on heating red oxide of mercury. Therefore, point 6 is completely altered in the final version of the Easter Memoir, and of course so is the final conclusion.

It is interesting to note how relatively few corrections and changes were required to make the original version of Lavoisier's paper correspond with the final and complete interpretation of his results. The similarity of the two versions is evidence that Lavoisier was very close to his final goal. Yet the first version is almost self-contradictory, for the gas from the red oxide is "common air" yet also "purer than common air." For the moment Lavoisier wanted to have it both ways! Though Lavoisier at first went wrong on an important point and was set right by Priestley, it was only Lavoisier, we must remember, who saw the real significance of the discovery of oxygen. He built the new facts into his own conceptual scheme and gave an enduring explanation of combustion. To Priestley goes the honor of being the effective discoverer of oxygen, but to the end of his life it was still "dephlogisticated air" to him.

In the following translation the new words and sentences *added* to the May 1775 version (Rozier's *Journal*) by Lavoisier are enclosed in braces and printed in italics. The material deleted from this first version in the preparation of the second and final form of the communication is printed in boldface type. If the changes are extensive the two versions are printed in parallel columns, the first version being on the left. The footnotes and the explanatory material enclosed in brackets were written by the translator, the editor of this Case, to assist the reader.

## 2. ENGLISH TRANSLATION OF THE TWO VERSIONS OF THE EASTER MEMOIR OF LAVOISIER

This memoir was read before the French Academy of Sciences at its Easter meeting in 1775, and again on August 8, 1778 (*Mémoires de l'Académie des Sciences,* 1775, p. 520). It carried the following footnote: "The first experiments relative to this Memoir were made more than a year ago; those on 'precipitate of mercury *per se*' were first attempted with a burning glass in the month of November, 1774, and continued with all precautions and necessary care in the laboratory of Montigny, conjointly with M. Trudaine on February 28 and March 1 and 2 of this year; finally they were once more repeated on the 31st of last March in the presence of M. The Duke de la Rochefoucault, and MM. Trudaine, de Montigny, Macquer and Cadet."

### *Memoir On The Nature Of The Principle Which Combines With Metals During Calcination and Increases Their Weight*

Are there different kinds of air? Does it suffice that a body should be in a state of permanent elasticity in order to be considered a kind of air? Are the different airs that nature offers us, or that we succeed in making, exceptional substances or modifications of atmospheric air? Such are the principal questions embraced in the plan I have formed and the problems which I propose to develop {*in succession*} before the Academy. But since the time devoted to our public meetings does not allow me to treat any one of these questions in full, I will confine myself today to one particular case, and will limit myself to showing that the principle which unites with metals during calcination, which increases their weight, and which is a constituent of the calx is

**neither one of the constituent parts of the air, nor a particular acid distributed in the atmosphere, that it is the air itself entire without alteration, without decomposition even to the point that if one sets it free after it has been so combined it comes out more pure, more respirable, if this expression may be permitted, than the air of the atmosphere and is more suitable to support ignition and combustion.**

{*nothing else than the healthiest and purest part of air; so that if air, after entering into combination with a metal, is set free again, it emerges in an eminently respirable condition, more suited than atmospheric air to support ignition and combustion.*}

The majority of metallic calces are only reduced, that is to say, only return to the metallic condition, by immediate contact with a carbonaceous material, or with some substance containing what is called phlogiston. The charcoal that one uses is entirely destroyed during the operation when the amount is in suitable proportion, whence it follows that the air set free from metallic reductions with charcoal is not simple; it is in some way the result of the combination of the elastic fluid set free from the metal and that set free from the charcoal; thus, though this fluid is obtained in the state of fixed air, one is not justified in concluding that it existed in this state in the metallic calx before its combination with the carbon.

These reflections have made me feel how essential it is, in order to unravel the mystery of the reduction of metallic calces, that all my experiments be performed on calces which can be reduced without any addition [that is, without the addition of charcoal]. The calces of iron have this property: in fact, all those, whether natural or artificial, which we have placed in the focus of a burning glass either that of M. the Regent or of M. de Trudaine, in each case were totally reduced without addition.

In consequence, I tried with the aid of a burning glass to reduce several kinds of iron calces under large glass bell-jars turned upside down in mercury, and by this method I succeeded in setting free a large quantity of **elastic fluid** {*air*}; but, as at the same time this **elastic fluid** {*air*} mixed with the common air contained in the bell, this circumstance cast great uncertainty on my results; none of the experiments to which I submitted this air was completely conclusive, and it was impossible to be sure whether the phenomena obtained depended on the common air, on that released from the iron calces, or on the combination of the two together. These experiments not having fulfilled my object, I omit the details here; they will find their natural place in other memoirs. [Those who have studied Case 1 of this series will be interested in comparing this method of reporting with that used by Boyle concerning his experiments with the subtle fluid.]

Since these difficulties proceeded from the very nature of iron, from the refractory quality of its calces, and from the difficulty of reducing them without addition, I considered them as insurmountable, and henceforth decided to address myself to another kind of calx more easily treated and which, like iron calces, possessed the property of being reduced without addition. *Mercurius calcinatus per se*, which is nothing else than a calx of mercury, as several authors have already maintained and as the reader of this Memoir will be even convinced, *mercurius calcinatus per se*, I repeat, seemed to me completely suitable for what I had in view; no one today is unaware that this substance can be reduced without addition at a very moderate degree of heat. Although I have repeated many times the experiments I am going to report, I have not thought it necessary to give particular details of any of them here, for fear of occupying too much space, and I have therefore combined into one account the observations made during several repetitions of the same experiment.

Throughout this memoir Lavoisier is referring to only one compound of mercury, namely the red oxide; he refers to it as *precipitated mercury per se,* to indicate that it was obtained by heating mercury by itself. We have translated it as *mercurius calcinatus per se,* the Latin phrase used by Priestley to describe the same material.

In order to be sure that *mercurius calcinatus per se* was a true metallic calx, whether it would give the same results and the same kind of air on reduction, I tried at first to reduce it by the ordinary method, that is to say, to use the accepted expression, by the addition of phlogiston. I mixed one ounce of this calx with 48 grains [3] of powdered charcoal, and introduced the whole into a little glass retort of at most two cubic inches capacity, which I placed in a reverberatory furnace of proportionate size. The neck of this retort was about a foot long, and three to four lines in diameter [a line is one-twelfth of an inch]; it had been bent in different places by means of an enameler's lamp, and its tip was such that it could be fixed under a bell-jar of sufficient size, filled with water, {*and turned upside down in a trough of water. The apparatus now before the eyes of the Academy suffices to give an idea of the operation.*} This apparatus, simple as it is, is the more exact that there is no soldering or luting or any passage through which air might leak in or escape.

As soon as a flame was applied to the retort and the heat had begun to take effect, the ordinary air contained in the retort expanded, and a small quantity passed into the bell-jar; but in view of the small size of the part of the retort that remained empty, this air could not introduce

[3] The system of weights employed was: 1 pound (Paris standard) = 16 ounces; 1 ounce = 8 gros; 1 gros = 72 grains; or 1 pound (Paris) = 9216 grains. The English avoirdupois pound of 16 ounces (still commonly used) is somewhat lighter, and is divided into 7000 grains.

a sensible error, and at the most it could scarcely amount to a cubic inch. When the retort began to get hotter, air was very rapidly disengaged and rose through the water into the bell-jar; the operation did not last more than three-quarters of an hour, the flame being used sparingly during this interval. When all the calx of mercury had been reduced and the air ceased rising, I marked the height at which the water stood in the bell-jar and found that the quantity of air set free amounted to 64 cubic inches, without allowing for the volume necessarily absorbed in passing through the water.

I submitted this air to a large number of tests, which I will not describe in detail, and found (1) that it was susceptible to combination with water on shaking and gave to it all the properties of acidulated, gaseous, or aerated waters such as those of Seltz, Pougues, Bussang, Pyrmont, etc.; (2) that animals placed in it died in a few seconds; (3) that candles and all burning bodies were instantly extinguished in it; (4) that it precipitated lime water; (5) that it combined very easily with either the fixed or the volatile alkalis, removing their causticity and giving them the property of crystallizing. [Fixed alkalis are materials like slaked lime or caustic soda that absorb carbon dioxide to form carbonates which, unlike the alkalis, are not caustic amorphous solids but noncaustic crystalline solids; volatile alkali is ammonia.] All these are precisely the qualities of the kind of air known as fixed air,

**or mephitic air such as is obtained from all metallic calces by the addition of charcoal** such as is set free from fermenting matters. It was thus unquestionable that *mercurius calcinatus per se* should be included in the category of metallic calces.

{*obtained by the reduction of minium* [an oxide of lead] *by powdered charcoal,*} such as is set free {*from calcareous earths and effervescent alkalis by their combination with acids, or*} from fermenting {*vegetable*} matters, {*etc.*}. It was thus unquestionable that *mercurius calcinatus per se* {*gave the same products as other metallic calces on reduction with addition of phlogiston and that it*} should {*consequently*} be included in the {*general*} category of metallic calces.

Lavoisier up to this point has added nothing new. Bayen, another French chemist, had recently reported similar results; Lavoisier's work represented a careful rechecking. How important such independent verification may be is evident from what follows. For Bayen had reported that the gas evolved when the mercury calx was heated by itself was also fixed air! How he could come to such an erroneous conclusion no one knows, but he was obviously not a skillful experimenter with gases.

It remained only to examine this calx alone, to reduce it without addition, to see if some **air** {*elastic fluid*} would be set free, and if so, to determine **what state it was in** {*its nature*}. To accomplish this I put into a retort of the same size as before (two cubic inches) one ounce of *mercurius calcinatus per se*, alone; I arranged the apparatus in the same way as for the preceding experiment, so that all the circumstances were exactly the same; the reduction was a little more difficult than with the addition of charcoal; it required more heat and there was no perceptible effect till the retort began to get slightly red-hot; then air was set free little by little, and passed into the bell-jar, and the same degree of heat being maintained for 2½ hours, all the *mercurius calcinatus per se* was reduced.

The operation completed, 7 gros, 18 grains of {*liquid*} mercury were found, partly in the neck of the retort and partly in a glass vessel which I had placed at the tip of the retort under the water; the amount of air in the bell-jar was found to be 78 cubic inches; from which it follows, that supposing all the loss of weight is attributed to the air, each cubic inch must weigh a little less than two-thirds of a grain, which does not differ much from the weight of ordinary air.

Having established these first results, I hastened to submit the 78 cubic inches of air I had obtained to all the tests suitable for determining its nature, and I found, much to my surprise:

| | |
|---|---|
| (1) that it was not susceptible to combination with water upon shaking; | (1) that it was not susceptible to combination with water upon shaking; |
| (2) that it did not precipitate lime water; | (2) that it did not precipitate lime water {*but only made it turbid to an almost imperceptible degree*}; |
| (3) that it did not combine with fixed or volatile alkalis; | (3) that it did not combine with fixed or volatile alkalis; |
| (4) that it did not at all diminish their caustic qualities; | (4) that it did not at all diminish their caustic qualities; |

[these first four tests were designed to show whether the gas was in whole or part "fixed air" as Bayen had reported; obviously it was not;]

| | |
|---|---|
| (5) that it could be used again for the calcination of metals; | (5) that it could be used again for the calcination of metals; |
| (6) **that it was diminished like common air by an addition of a third of nitrous air;** [4] | |

[4] "Addition of a third of nitrous air" means that one volume of "nitrous air" to two volumes of air to be tested was used — the mixture being, of course, allowed to stand over water.

finally, that it had none of the properties of fixed air: far from causing animals to perish, it seemed on the contrary more suited to support their respiration; not only were candles and burning objects not extinguished in it, but the flame increased in a very remarkable manner and gave much more light than in common air.

All these circumstances convinced me fully that this air **was not only common air but that it was more respirable, more combustible,** and consequently that it was more pure than even the air in which we live.

(6) finally, that it had none of the properties of fixed air: far from causing animals to perish, it seemed on the contrary more suited to support their respiration; not only were candles and burning objects not extinguished in it, but the flame increased in a very remarkable manner and gave much more light than in common air; {*charcoal burned in it with a radiance almost like that of phosphorus, and all combustible bodies in general were consumed with astonishing speed.*} All these circumstances convinced me fully that this air, {*far from being fixed air, was in a state more respirable and more combustible,*} and consequently that it was more pure than even the air in which we live.

It seems proved accordingly that the principle which combines with metals during calcination and which augments their weight is nothing else than the purest part of the very air which surrounds us, which we breathe, and which in this operation passes from a condition of elasticity to that of solidity; if then it is obtained in the form of fixed air in all the metallic reductions in which charcoal is used, this effect is due to **the charcoal itself** {*the combination of the latter with the pure part of the air*}, and it is very likely that all metallic calces, like that of mercury, would give **only common air** {*only eminently respirable air*} if they could all be reduced without addition like *mercurius calcinatus per se.*

The nitre referred to in the next paragraph is potassium nitrate, a compound of the metal potassium, oxygen, and nitrogen; from it chemists early learned to prepare nitric acid and hence other nitrates. Mixed with charcoal and sulfur it yields black gunpowder. This paragraph, which deals with the explosion of powder, may be omitted.

Everything that has been said of the air from metallic calces may naturally be applied to that obtained from nitre by detonation; we know from a number of experiments already published, the majority of which I have repeated, that most of this air is in the form of fixed air, that it is fatal to any animal breathing it, that it has the property of **precipitating lime water and of** combining with lime and alkalis, of softening them and making them crystallize; but since the detonation of nitre only takes place after the addition of charcoal or of some body which contains phlogiston, **it is very probable** {*it cannot be doubted*} that in this circumstance also **common air** {*eminently respirable air*} is converted into fixed air;

from which it follows that the air combined with nitre which produces the terrible explosions of gunpowder is **common** atmospheric air deprived of its elasticity. **From the fact that common air changes to fixed air when combined with charcoal it would seem natural to conclude that fixed air is nothing but a combination of common air and phlogiston. This is Mr. Priestley's opinion and it must be admitted that it is not without probability; however, when one looks into the facts in detail, contradictions arise so frequently I feel it necessary to ask natural philosophers and chemists still to suspend judgment; I hope to be soon in a position to communicate the reasons for my doubts.**

from which it follows that the air combined with nitre which produces the terrible explosions of gunpowder is {*the respirable part of*} atmospheric air deprived of its elasticity {*and which is one of the constituent principles of nitric acid. Since charcoal disappears completely in the revivification of mercuric calx and since only mercury and fixed air are produced by this operation, one is forced to conclude that the principle to which has hitherto been given the name of fixed air, is the result of the combination of the eminently respirable portion of the air with charcoal; and this is what I propose to develop in a more satisfactory manner in subsequent memoirs which I shall devote to this object.*}

A diagram of the apparatus used by Lavoisier in reducing red oxide of mercury with and without the addition of charcoal is shown in Fig. 1. The use of a reverberatory furnace (that is, one in which the heat is reflected down on the retort as well as coming from the flame below) enabled Lavoisier to heat the retort to a dull red heat. This temperature is required for the decomposition of the oxide if no charcoal is employed. Lavoisier does not state in this paper how he removed samples of the gas for the various tests, but in some of his later equipment the gas was removed through a small orifice near the top of the bell jar to smaller vessels by a suitable suction device. The methods of collecting gases over water or mercury and transferring them from one vessel to another had been greatly improved by Joseph Priestley a short time before.

The paragraph on page 26 in which Lavoisier records the weight of mercury recovered from the decomposition of the oxide and the amount of gas evolved is of great significance. Here Lavoisier is invoking a principle that underlay his whole work. One biographer of the French chemist, recalling that Lavoisier was a man of business affairs first and a chemist second, has spoken of Lavoisier's "principle of the balance sheet." Lavoisier had early in his career used the balance, and in repeating Joseph Black's work had undoubtedly been impressed by the Scotch chemist's success in accounting for the weights of the starting materials and of the products in a chemical reaction. In the

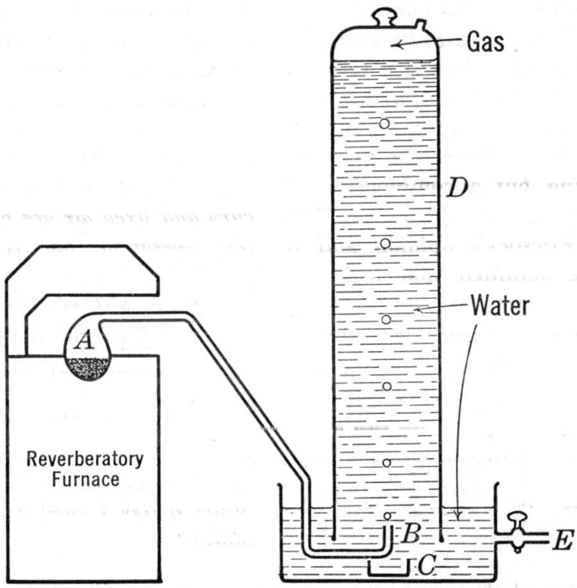


FIG. 1. Diagram of apparatus used by Lavoisier in collecting the gas evolved when red oxide of mercury is heated.

paragraph mentioned we see Lavoisier in his first major chemical contribution comparing the weight of the initial oxide (1 ounce) with the mercury produced (7 gros 18 grains) and deducing that the only other product (the gas) must account for the difference in weight (1 ounce = 8 gros and 1 gros = 72 grains; therefore the difference would be 7 gros 72 grains minus 7 gros 18 grains = 54 grains). He concludes that the 78 cubic inches of the air evolved must weigh 54 grains or about 54/78 grain per cubic inch (54/78 is a little more than 2/3, which is 52/78).

The weight of 1 cubic inch of air at the usual temperature and pressure of a laboratory had been given as 0.46 (or 36/78) grain by Lavoisier in a book published a few years earlier. Therefore, his statement that the estimated weight of 2/3 grain per cubic inch "does not differ much from the weight of ordinary air" is an optimistic statement by an investigator looking for an agreement between two numbers. Lavoisier was aware, of course, that this estimate of the density of the air was extremely rough, for his method of insuring the collection of all the mercury was far from sufficient for the purposes. (Note that the weight of the air evolved was a small difference between two relatively large numbers; such a situation is always unfavorable for accurate measurements. The failure to collect 3 percent, or 16 grains, of the mercury would have increased the estimated weight of the total gas by that amount, or by 16/78 grain per cubic inch, which is about the discrepancy observed.) Lavoisier does not record the pressure or temperature, though he does so in all his later experiments, where his calculations can be made with a higher degree of accuracy. The significance of the paragraph lies, therefore, not in the result, which is far too approximate to be useful, but in Lavoisier's mode of reasoning. Within the limits of experimental error he is trying to apply the *principle of the balance sheet.*

In his textbook of 1789 Lavoisier wrote: "We must lay it down as an incontestable axiom, that in all the operations of art and nature, nothing is created; an equal quantity of matter exists both before and after the experiment . . . Upon this principle, the whole art of performing chemical experiments depends." Whether or not Lavoisier came to this principle because as a member of the firm that collected taxes he was used to balancing his books in terms of money, the principle was the foundation of his work. In the nineteenth century the exactness of this principle was tested by very careful experimentation, using balances far more sensitive than those available in Lavoisier's time. In every case it was found that the weight of the factors was equal to that of the products within the small experimental error of the measurements. The principle was thus considered a generalization of experimental fact, rather than an axiom; it became known as the Law of Conservation of Mass (or Matter). We now have good reason to believe on both theoretical and experimental grounds that in the form in which it was expressed by Lavoisier and by the scientists of the nineteenth century it is only an approximation. Whenever energy is released or absorbed in a chemical change (and that is in practically every instance) a very very small change in mass occurs; the change is too small to be detected even with the most accurate measurements of the weights involved in even the most energetic chemical reaction. Those changes, however, that take place when the nuclei of atoms are altered liberate amounts of energy of another order of magnitude from those connected with chemical reactions; in this case the change of mass is significant and is directly related to the energy evolved. It is common nowadays to speak of the combination in the twentieth century of the two principles of conservation of mass and of conservation of energy into one — the principle of conservation of mass and energy.

This doctrine is accepted as a guiding principle by physicists and chemists today.

Those who have some knowledge of quantitative chemistry will interpret the equations given on page 13 for the reduction of mercury oxide with and without charcoal as stating that the volumes of oxygen and of carbon dioxide produced from 1 ounce of oxide should be the same (at the same temperature and pressure). Actually Lavoisier obtained 64 cubic inches of carbon dioxide and 78 of oxygen; the difference is due to the solubility of carbon dioxide in water. Indeed, if the gas had been allowed to stand over the water for any length of time or had been shaken with it, almost all of it would have gone into solution. In later experiments Lavoisier collected carbon dioxide over mercury. In this memoir we see him struggling with the difficulties of carrying out chemical experiments in such a way as to make quantitative observations meaningful. This is the beginning of the long and arduous road that had to be traveled before chemistry could become a quantitative science.

Lavoisier's reasoning in 1775 and 1778 is worth examining with some care. His broad working hypothesis of 1772 was that "something" was taken up from the air when a metal was calcined. At first he thought that this "something" might be fixed air and had experimented in vain to prove this. In other words, his broad working hypothesis, when made specific by substituting the words "fixed air" for "something," yielded deductions that were *not* confirmed by experimental test. Now, thanks perhaps to the unconscious tip that Priestley had conveyed to him in conversation in the fall of 1774, he had a calx which on heating yielded a gas that behaved like common air. Substituting common air for the "something" in his broad working hypothesis yielded a deduction that appeared to be confirmed by the experiment. But note carefully that deductions from broad working hypotheses are never directly confirmed or negated. A specific experiment must always be related to the deductions by one or more limited working hypotheses. This is where the difficulties arise and in this case we can see exactly how they arose in 1775. Of the six tests recorded by Lavoisier in his first version of the Easter Memoir, the first four gave convincing evidence that the gas was not fixed air. (In each test a limited working hypothesis was implicit, an "if . . . then" type of reasoning was employed.) The fifth and sixth tests, together with the experiments with the candle and with animals, seemed to provide conclusive evidence that the gas was common air. On the supposition that the gas was common air, Lavoisier could say, "If I perform the following manipulations, then the result will be such and such." (This last is a limited working hypothesis that is confirmed or negated by test.) In other words, there is the haunting question whether another substance could also behave in this same manner; of these tests the nitrous air test appeared to be the most specific and must have appealed to Lavoisier because it was at least roughly quantitative. But the similarity in behavior of the new gas and common air in this test was, we have seen, only apparent. The assumption that there would be no further diminution in volume if more nitrous air were added was implicit in the report of Easter 1774, for that was the case with common air. Priestley, as we shall see in the next section, corrected the assumption as a result of an accident; Lavoisier corrected it after having learned of Priestley's use of the nitrous air test to show that the gas was *not* common air.

As far as the significance of the nitrous air test was concerned, both Priestley and Lavoisier agreed by the spring of 1776. They would now say that an air is common air if it is diminished by a third of nitrous air, *provided* it is not further diminished on adding more nitrous air. They would both agree that a new air was present when the calx of mercury was heated. But as to the broad working hypothesis of Lavoisier, they disagreed completely. Priestley stuck to the conceptual scheme in which phlogiston was the determining factor in calx formation. Lavoisier saw his broad working hypothesis now made specific by substituting the words "a constituent of the atmosphere which supports combustion" for his "something." His working hypothesis on a grand scale was about to attain the status of a new conceptual scheme.

### 3. PRIESTLEY'S COMMENTS ON LAVOISIER'S EASTER MEMOIR

Priestley's comments on Lavoisier's paper were printed in Section XVI of Volume II of his *Experiments and Observations on Different Kinds of Air* the dedication of which is dated November 1, 1775. The title of this section, one of the last in the book, is "An Account of some Misrepresentations of the Author's Sentiments, and of some Differences of Opinion with respect to the Subject of Air." The relevant passages (from pages 320–323) follow [5] (the words included in brackets have been added to enable the reader to follow the discussion in modern terms); the opening words refer to a discussion of Lavoisier's remark about Priestley's opinion as to the relation of fixed air to phlogiston, which is not relevant to the discussion of the experiments.

> Having mentioned the paper of Mr. Lavoisier's, published in Mr. Rosier's Journal, I would observe, that it appears by it, that, after I left Paris, where I procured the *mercurius calcinatus* [red oxide of mercury] above mentioned, and had spoken of the experiments that I had made, and that I intended to make with it, he began his experiments upon the

[5] All extracts from Priestley's book used in this Case are taken from the second edition (1776).

same substance, and presently found what I have called *dephlogisticated air* [oxygen], but without investigating the nature of it, and indeed, without being fully apprised of the degree of its purity. For he had only tried it with one-third of nitrous air, and observed that a candle burned in it with more vigour than in common air; and though he says it *seems to be* more fit for respiration than common air, he does not say that he had made any trial how long an animal could live in it.

Priestley felt he had been unfairly treated by Lavoisier because the French chemist made no mention in his paper of the facts communicated to him in the fall of 1774 by Priestley. Historians have been discussing ever since how essential this verbal communication was in fact. The incident, quite apart from the entertaining way in which Priestley rebukes Lavoisier, is of interest as illustrating the state of scientific communication toward the end of the eighteenth century. As in the seventeenth century, investigators did not feel under any special obligation to refer to the work of others. Lavoisier makes no mention of Bayen's paper published a year before; possibly he assumed that everybody had read it. Since the middle of the nineteenth century, practice has changed in this respect. Investigators are today very scrupulous in referring to published work and even verbal communications.

He therefore inferred, as I have said that I myself had once done, that this substance had, during the process of calcination, imbibed atmospherical air, not in part, but in whole. But then he extends his conclusion, and, as it appears to me, without any evidence, to all the metallic calces; saying that, very probably, they would all of them yield only common air, if, like *mercurius calcinatus*, they could be reduced without addition. For he considers the fixed air [carbon dioxide], which is yielded by most of them, to come from the charcoal, made use of to revivify the calx. Whereas it will be seen, in the course of my experiments, that several of those calces yield fixed air by *heat only*, without any addition of charcoal [that is, without addition of charcoal before being heated. Priestley was in error here; Lavoisier was right. Those calces from which Priestley obtained fixed air (carbon dioxide) were impure; they contained carbonates which are decomposed on heating yielding carbon dioxide].

He adds, that since common air is changed into fixed air when it is combined with charcoal, it would seem natural to conclude, that fixed air is only a combination of common air and phlogiston (an opinion which, as has been seen before, he ascribes to me) and it is not, he says, without probability; but adds, that it is so often contradicted by facts, that he desires philosophers and chymists to suspend their judgments; hoping that it will soon be in his power to explain the motives of his doubts. I, for one, am waiting with some impatience for this explanation. [But when it came, Priestley rejected it!]

Mr. Lavoisier also concludes, from his observations, that the air produced by the detonation of nitre and the firing of gunpowder is common air. When he sees this volume of mine, he will, I doubt not, be convinced of the imperfection of his theory, and of this mistake, which he has been led into by means of it. [The detonation of nitre is not relevant to the case at hand.]

Mr. Lavoisier as well as Sig. Landriani, Sig. F. Fontana, and indeed all other writers except myself, seems to consider common air (divested of the effluvia that float in it, and various substances that are dissolved in it, but which are in reality foreign to it) as a simple *elementary body*; whereas I have, for a long time, considered it as a *compound*; and this notion has been of great service to me in my inquiries.

As a concurrence of unforeseen and undesigned circumstances has favoured me in this inquiry, a like happy concurrence may favour Mr. Lavoisier in another; and as, in this case, truth has been the means of leading him into error, error may, in its turn, lead him into truth. It will have been seen, in the course of my writings, that both these circumstances have frequently happened to myself; and indeed examples of both of them will be found in my first section concerning this very subject of dephlogisticated air.

It is pleasant when we can be equally amused with our own mistakes, and those of others. I have voluntarily given others many opportunities of amusing themselves with mine, when it was entirely in my power to have concealed them. But I was determined to shew how little *mystery* there really is in the business of experimental philosophy, and with how little *sagacity*, or even *design*, discoveries (which some persons are pleased to consider as great and wonderful things) have been made.

### 4. THE STEPS LEADING TO PRIESTLEY'S DISCOVERY OF OXYGEN

Having read Priestley's comments written in the summer of 1775 on Lavoisier's Easter Memoir of the same year, we are ready to examine the record of how Priestley himself at first erroneously interpreted his experiments with oxygen. In the summer of 1774 he first started studying the gas evolved when red oxide of mercury is heated, as he relates in the account reproduced here.

A few words of explanation are required to make clear to the reader who has little or no knowledge of chemistry how Priestley made his first error, which was a completely erroneous identification of the new gas. He tested the "new air" (which was in reality oxygen) with a lighted candle and found that the candle was not extinguished but burned brightly. Now a few years before, Priestley had prepared an oxide of nitrogen that has the unusual property of supporting combustion. He did not know, of course, that it was an oxide of nitrogen, but he called it "phlogisticated nitrous air" for he had prepared it by letting "nitrous air" (nitric oxide) stand with metallic iron or certain other substances which decompose the nitric oxide, taking away some of the oxygen and yielding the other oxide, which supports combustion. Although in Priestley's terminology it was "phlogisticated," he apparently gave no explanation of why phlogistication made the nitrous air inflammable. The chemical reaction is as follows:

| | | | slowly | | | |
|---|---|---|---|---|---|---|
| $2NO$ | $+$ | $Fe$ | $\rightarrow$ | $N_2O$ | $+$ | $FeO$ |
| "nitrous air," colorless, insoluble gas; will not support combustion | plus | metallic iron | yields | "phlogisticated nitrous air" (modern name: nitrous oxide) somewhat soluble in water; supports combustion; commonly called "laughing gas" | plus | iron oxide |

The product we now call nitrous oxide; this modern name is confusing in this narrative, since in Priestley's day our modern nitric oxide was called "nitrous air." Fortunately, nitrous oxide plays only a minor role and will hereafter be designated as "laughing gas," for it is often so called from the fact that today it is used for anesthesia, particularly by dentists, and the patient sometimes becomes quite merry before the anesthesia is complete! Priestley's sample of this gas was a mixture of the original nitric oxide and "laughing gas." When the mixture was agitated with water, the oxide that supports combustion was dissolved. The significance of this fact will be evident in what follows.

Priestley's account of his experiments leading to the identification of oxygen as a new gas are described in Section III of Volume II of his *Experiments*. After some rather amusing general remarks about experimental difficulties, he recounts the experiments performed in 1774 and the totally erroneous conclusions he drew from them. (It was at this time that he confused oxygen with "phlogisticated nitrous air," that is, "nitrous air exposed to iron.") On March 1, 1775 Priestley decided to apply his "nitrous air test" to the "air" obtained from the red oxide. The results of this experiment showed that the new "air" was at least very much like common air. It is a remarkable coincidence that Lavoisier almost simultaneously and quite independently was performing the same experiment; he decided on the basis of the result that he had at hand a purified common air, as is evident from the first version of his memoir. But Priestley, more or less by accident, went further. He found that the residual gas, or air, left after the reaction with nitrous air and the subsequent diminution in volume would still allow a candle to burn brightly. This accidental discovery was followed up by experiments with a mouse which, he found, would live in this residual gas. Priestley then began to suspect that the air under investigation not only was *at least as good as common air* but might be *better*. This he proved by showing that even after a mouse had lived in it for some time, the nitrous air diminished its volume more than common air.

"Being now fully satisfied of the superior goodness of this kind of air," he found by careful trials with increasing amounts of nitrous air that he could get an increasing diminution in volume until more than four half measures (for one measure of the gas being tested) were used. This proved that a substance quite different from common air was at hand. The effective discovery of oxygen by Priestley had occurred.

Priestley's own account now follows (pages 29–49 of Volume II). The material in brackets is added to assist the reader.

### General Observations on the Role of Chance Discoveries

The contents of this section will furnish a very striking illustration of the truth of a remark, which I have more than once made in my philosophical writings, and which can hardly be too often repeated, as it tends greatly to encourage philosophical investigations; viz. that more is owing to what we call *chance*, that is, philosophically speaking, to the observation of *events arising from unknown causes*, than to any proper *design*, or pre-conceived *theory* in this business. This does not appear in the works of those who write *synthetically* upon these subjects; but would, I doubt not, appear very strikingly in those who are the most celebrated for their philosophical acumen, did they write *analytically* and ingenuously.

For my own part, I will frankly acknowledge, that, at the commencement of the experiments recited in this section, I was so far from having formed any hypothesis that led to the discoveries I made in pursuing them, that they would have appeared very improbable to me had I been told of them; and when the decisive facts did at length obtrude themselves upon my notice, it was very slowly, and with great hesitation, that I yielded to the evidence of my senses. And yet, when I re-consider the matter, and compare my last discoveries relating to the constitution of the atmosphere with the first, I see the closest and the easiest connexion in the world between them, so as to wonder that I should not have been led immediately from the one to the other. That this was not the case, I attribute to the force of prejudice, which, unknown to ourselves, biasses not only our *judgments*, properly so called, but even the perceptions of our senses: for we may take a maxim so strongly for granted, that the plainest evidence of sense will not intirely change, and often hardly modify our persuasions; and the more ingenious a man is, the more effectually he is entangled in his errors; his ingenuity only helping him to deceive himself, by evading the force of truth.

These observations of Priestley's have been the basis for some criticism of his scientific judgment by historians of science. As in some of the comments on the accidental nature of his experiments later in this same account, Priestley is here frank almost to the point of naïveté. Yet anyone conversant with the way chemical discoveries are made even today will hardly take exception to his remarks.

## Priestley's Views on the Nature of the Atmosphere

In the following passage, Priestley refers to "agitation in water" as a method of purifying a gas. In this process, for which he had invented an apparatus, two things happened of which he was ignorant and both of which made the gas more fit for respiration. One was the dissolving in the water of the carbon dioxide in the gas; the other was the addition to the gas of some oxygen, which came from the dissolved air present in ordinary samples of water. Priestley is quite correct in what he says about the "process of vegetation"; green plants not only use up carbon dioxide but evolve oxygen.

> There are, I believe, very few maxims in philosophy that have laid firmer hold upon the mind, than that air, meaning atmospherical air (free from various foreign matters, which were always supposed to be dissolved, and intermixed with it) is a *simple elementary substance*, indestructible, and unalterable, at least as much so as water is supposed to be. In the course of my inquiries, I was, however, soon satisfied that atmospherical air is not an unalterable thing; for that the phlogiston with which it becomes loaded from bodies burning in it, and animals breathing it, and various other chemical processes, so far alters and depraves it, as to render it altogether unfit for inflammation, respiration, and other purposes to which it is subservient; and I had discovered that agitation in water, the process of vegetation, and probably other natural processes, by taking out the superfluous phlogiston, restore it to its original purity. But I own I had no idea of the possibility of going any farther in this way, and thereby procuring air purer than the best common air. I might, indeed, have naturally imagined that such would be air that should contain less phlogiston than the air of the atmosphere; but I had no idea that such a composition was possible. [Here as throughout his life Priestley is using the concept of phlogiston to explain combustion and calcination.]

## Priestley's First Experiments with Red Oxide of Mercury

Publication of these results in Priestley's *Experiments*, as given below, followed the communication of three letters announcing the essential facts to the Royal Society; they are dated March 15, April 1, and May 29, 1775, and were printed in volume 65 of the *Philosophical Transactions of the Royal Society* for that year, which seems to have appeared more promptly than the corresponding record of the French Academy.

> At the time of my former publication, I was not possessed of a *burning lens* of any considerable force; and for want of one, I could not possibly make many of the experiments that I had projected, and which, in theory, appeared very promising. I had, indeed, a *mirror* of force sufficient for my purpose. But the nature of this instrument is such, that it cannot be applied, with effect, except upon substances that are capable of being suspended, or resting on a very slender support. It cannot be directed at all upon any substance in the form of *powder*, nor hardly upon any thing that requires to be put into a vessel of quicksilver [liquid metallic mercury]; which appears to me to be the most accurate method of extracting air from a great variety of substances, as was explained in the Introduction to this volume. But having afterwards procured a lens of twelve inches diameter, and twenty inches focal distance, I proceeded with great alacrity to examine, by the help of it, what kind of air a great variety of substances, natural and factitious, would yield, putting them into the vessels . . . which I filled with quicksilver, and kept inverted in a bason of the same. Mr. Warltire, a good chymist, and lecturer in natural philosophy, happening to be at that time in Calne, I explained my views to him, and was furnished by him with many substances, which I could not otherwise have procured.

Priestley's statements about the mirror are at first sight somewhat perplexing. The point is not important, however. Apparently Priestley had a concave mirror enabling him to bring a beam of sunlight to a focus and thus concentrate the radiant energy. However, the material placed at the focus of a concave mirror is *between* the source of light and the reflecting surface. It must be small if it is not to cut off a large portion of the beam of light. With a lens there is, of course, no such difficulty (see Fig. 2).

> With this apparatus, after a variety of other experiments, an account of which will be found in its proper place, on the 1st of August, 1774, I endeavoured to extract air from *mercurius calcinatus per se*; and I presently found that, by means of this lens, air was expelled from it very readily. [The gas evolved, which was oxygen, displaced the mercury.] Having got about three or four times as much as the bulk of my materials, I admitted water to it, and found that it was not imbibed by it. [This showed that the gas was not carbon dioxide which he suspected would be formed.] But what surprized me more than I can well express, was, that a candle burned in this air with a remarkably vigorous flame, very much like that enlarged flame with which a candle burns in nitrous air, exposed to iron or liver of sulphur [calcium sulfide which, like iron,

transforms nitric oxide into laughing gas (see p. 35); that is, Priestley thought he was dealing with what we now call laughing gas]; but as I had got nothing like this remarkable appearance from any kind of air besides this particular modification of nitrous air [laughing gas], and I knew no nitrous acid was used in the preparation of *mercurius calcinatus*, I was utterly at a loss how to account for it. ["Nitrous air" requires a nitrate for its preparation and this in turn requires nitric acid or its equivalent. This acid was called nitrous acid in Priestley's day.]

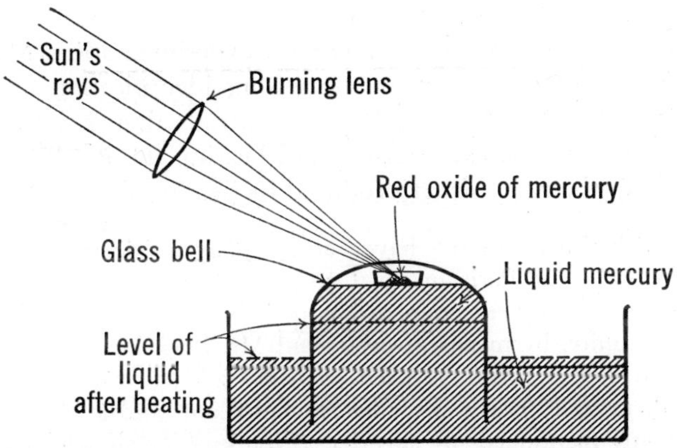


FIG. 2. Priestley's method of heating red oxide of mercury with a burning lens.

In this case, also, though I did not give sufficient attention to the circumstance at that time, the flame of the candle, besides being larger, burned with more splendor and heat than in that species of nitrous air [i.e., laughing gas]; and a piece of red-hot wood sparkled in it, exactly like paper dipped in a solution of nitre [potassium nitrate used in the manufacture of gunpowder], and it consumed very fast; an experiment which I had never thought of trying with nitrous air. [Priestley probably means here nitrous air modified by exposure to iron.]

At the same time that I made the above-mentioned experiment, I extracted a quantity of air, with the very same property, from the common *red precipitate*, which being produced by a solution of mercury in spirit of nitre [nitric acid], made me conclude that this peculiar property, being similar to that of the modification of nitrous air above mentioned, depended upon something being communicated to it by the nitrous acid [nitric acid]; and since the *mercurius calcinatus* is produced by exposing mercury to a certain degree of heat, where common air has access to it, I likewise concluded that this substance had collected something of *nitre*, in that state of heat, from the atmosphere [a not unlikely supposition but erroneous, for it is in fact the oxygen of the nitric acid that is left united with the mercury when the red oxide is prepared from the nitrate].

This, however, appearing to me much more extraordinary than it ought to have done, I entertained some suspicion that the mercurius calcinatus, on which I had made my experiments, being bought at a common apothecary's, might, in fact, be nothing more than red precipitate; though, had I been any thing of a practical chymist, I could not have entertained any such suspicion. However, mentioning this suspicion to Mr. Warltire, he furnished me with some that he had kept for a specimen of the preparation, and which, he told me, he could warrant to be genuine. This being treated in the same manner as the former, only by a longer continuance of heat, I extracted much more air from it than from the other.

This is an example of careful chemical experimentation. More than one chemist has gone astray by the wrong identification of his materials. Red oxide prepared via mercury nitrate, that is, by dissolving mercury in nitric acid, evaporating, and heating the residue might be different from the red powder produced by heating mercury in air.

This experiment might have satisfied any moderate skeptic: but, however, being at Paris in the October following, and knowing that there were several very eminent chymists in that place, I did not omit the opportunity, by means of my friend Mr. Magellan, to get an ounce of mercurius calcinatus prepared by Mr. Cadet, of the genuineness of which there could not possibly be any suspicion; and at the same time, I frequently mentioned my surprize at the kind of air which I had got from this preparation to Mr. Lavoisier, Mr. le Roy, and several other philosophers, who honoured me with their notice in that city; and who, I dare say, cannot fail to recollect the circumstance.

By an interesting coincidence Lavoisier obtained his sample of red oxide of mercury from the same Parisian apothecary. The two samples may well have come from the same bottle. Priestley's conversation with Lavoisier has been the subject of much debate among historians of science. Since at this time he was inclined to think the gas in question was laughing gas, he could not have told Lavoisier of the discovery of a new gas. Yet the fact that the gas was *not* carbon dioxide ("fixed air") and that it did support combustion was an important fact to one who, like Lavoisier, was intent on solving the problem of combustion and calcination. The French chemist Bayen in February of the same year had published the erroneous statement that the red powder on heating alone gave fixed air. Bayen's work may have provided the stimulus for Lavoisier's examination of the behavior of red oxide of mercury on heating, but it is far more likely that it was Priestley's remarks to him. These were made in October; according to Lavoisier's own statement, he started his experiments in November.

At the same time, I had no suspicion that the air which I had got from the mercurius calcinatus was even wholesome, so far was I from knowing what it was that I had really found; taking it for granted, that it was nothing more than such kind of air as I had brought nitrous air to be by the processes above mentioned; and in this air I have observed that a candle would burn sometimes quite naturally, and sometimes with a beautiful enlarged flame, and yet remain perfectly noxious.

At the same time that I had got the air above mentioned from mercurius calcinatus and the red precipitate, I had got the same kind from *red lead* or *minium* [a red oxide of lead]. In this process, that part of the minium on which the focus of the lens had fallen, turned yellow. One third of the air, in this experiment, was readily absorbed by water [due to some carbon dioxide from carbonate mixed with the oxide], but, in the remainder, a candle burned very strongly, and with a crackling noise.

That fixed air [carbon dioxide] is contained in red lead I had observed before; for I had expelled it by the heat of a candle, and had found it to be very pure. See Vol. I, p. 192. I imagine it requires more heat than I then used to expel any of the other kind of air. [This is correct; the carbonate decomposes easily. However, Priestley assumes that his red lead is a homogeneous substance. His sample clearly was not, but contained some lead carbonate. One of the advantages of mercury oxide for preparing oxygen is that unlike many other oxides it is not contaminated with carbonate by any method of preparation.]

This experiment with *red lead* confirmed me more in my suspicion, that the *mercurius calcinatus* must get the property of yielding this kind of air from the atmosphere, the process by which that preparation, and this of red lead is made, being similar. As I never make the least secret of any thing that I observe, I mentioned this experiment also, as well as those with the mercurius calcinatus, and the red precipitate, to all my philosophical acquaintance at Paris, and elsewhere; having no idea, at that time, to what these remarkable facts would lead.

Presently after my return from abroad, I went to work upon the *mercurius calcinatus*, which I had procured from Mr. Cadet; and, with a very moderate degree of heat, I got from about one fourth of an ounce of it, an ounce-measure of air [the relation of weight to volume of air is of no particular significance], which I observed to be not readily imbibed, either by the substance itself from which it had been expelled (for I suffered them to continue a long time together before I transferred the air to any other place) or by water, in which I suffered this air to stand a considerable time before I made any experiment upon it.

In this air, as I had expected, a candle burned with a vivid flame; but what I observed new at this time, (Nov. 19,) and which surprized me no less than the fact I had discovered before, was, that, whereas a few moments agitation in water will deprive the modified nitrous air of its property of admitting a candle to burn in it; yet, after more than ten times as much agitation as would be sufficient to produce this alteration in the nitrous air, no sensible change was produced in this. A candle still burned in it with a strong flame. . . [The "nitrous air" here mentioned is nitrous oxide or laughing gas which, as already stated, is soluble in water. Priestley's preparation of this gas from nitric oxide resulted in a mixture; agitation with water removed the nitrous oxide.]

But I was much more surprized, when, after two days, in which this air had continued in contact with water (by which it was diminished about one twentieth of its bulk) I agitated it violently in water about five minutes, and found that a candle still burned in it as well as in common air. The same degree of agitation would have made phlogisticated nitrous air [laughing gas] fit for respiration indeed, but it would certainly have extinguished a candle [because most of the laughing gas would have been dissolved in the water; the difference in solubility of laughing gas and oxygen in water gave Priestley his first hint].

These facts fully convinced me, that there must be a very material difference between the constitution of the air from mercurius calcinatus, and that of phlogisticated nitrous air [laughing gas], notwithstanding their resemblance in some particulars. But though I did not doubt that the air from *mercurius calcinatus* was fit for respiration, after being agitated in water, as every kind of air without exception, on which I had tried the experiment, had been, I still did not suspect that it was respirable in the first instance; so far was I from having any idea of this air being, what it really was, much superior, in this respect, to the air of the atmosphere.

In this ignorance of the real nature of this kind of air, I continued from this time (November) to the 1st of March following; having, in the mean time, been intent upon my experiments on the vitriolic acid air above recited, and the various modifications of air produced by spirit of nitre, an account of which will follow. [These experiments are not relevant.] But in the course of this month, I not only ascertained the nature of this kind of air, though very gradually, but was led by it to the complete discovery of the constitution of the air we breathe.

Till this 1st of March, 1775, I had so little suspicion of the air from mercurius calcinatus, &c. being wholesome, that I had not even thought of applying to it the test of nitrous air [on this point Lavoisier was quicker on the trigger, for he applied this test at once, though, as we have seen, with unfortunate results]; but thinking (as my reader must imagine I frequently must have done) on the candle burning in it after long agitation in water, it occurred to me at last to make the experiment; and putting one measure of nitrous air to two measures of this air [the standard procedure devised by Priestley himself], I found, not only that it was diminished, but that it was diminished quite as much as common air [Priestley misses the quantitative difference at this time; see the next paragraph], and that the redness of the mixture was likewise equal to that of a similar mixture of nitrous and common air.

After this I had no doubt but that the air from mercurius calcinatus was fit for respiration, and that it had all the other properties of genuine common air. But I did not take notice of what I might have observed, if I had not been so fully possessed by the notion of there being no air better than common air, that the redness was really deeper, and the diminution something greater than common air would have admitted. [See p. 19 for explanation.]

Moreover, this advance in the way of truth, in reality, threw me back into error, making me give up the hypothesis I had first formed, viz. that the mercurius calcinatus had extracted spirit of nitre from the air; for I now concluded, that all the constituent parts of the air were equally, and in their proper proportion, imbibed in the preparation of this substance, and also in the process of making red lead. [As far as errors are concerned there is little to choose; both hypotheses, in fact, were wrong!] For at the same time that I made the above-mentioned experiment on the air from mercurius calcinatus, I likewise observed that the air which I had extracted from red lead, after the fixed air was washed out of it, was of the same nature, being diminished by nitrous air like common air: but, at the same time, I was puzzled to find that air from the red precipitate was diminished in the same manner, though the process for making this substance is quite different from that of making the two others. But to this circumstance I happened not to give much attention.

I wish my reader be not quite tired with the frequent repetition of the word *surprize*, and others of similar import; but I must go on in that style a little longer. For the next day I was more surprized than ever I had been before, with finding that, after the above-mentioned mixture of nitrous air and the air from mercurius calcinatus, had stood all night, (in which time the whole diminution must have taken place; and, consequently, had it been common air, it must have been made perfectly noxious, and intirely unfit for respiration or inflammation [i.e., would have been nitrogen]) a candle burned in it, and even better than in common air.

I cannot, at this distance of time, recollect what it was that I had in view in making this experiment; but I know I had no expectation of the real issue of it. Having acquired a considerable degree of readiness in making experiments of this kind, a very slight and evanescent motive would be sufficient to induce me to do it. If, however, I had not happened, for some other purpose, to have had a lighted candle before me, I should probably never have made the trial; and the whole train of my future experiments relating to this kind of air might have been prevented. [Priestley's critics have made much fun of this honest statement. Few experimenters are as frank about their lucky strikes.]

Still, however, having no conception of the real cause of this phenomenon, I considered it as something very extraordinary; but as a property that was peculiar to air extracted from these substances, and *adventitious*; and I always spoke of the air to my acquaintance as being substantially the same thing with common air. I particularly remember my telling Dr. Price, that I was myself perfectly satisfied of its being common air, as it appeared to be so by the test of nitrous air; though, for the satisfaction of others, I wanted a mouse to make the proof quite complete.

Up to this point, Priestley is just where Lavoisier was, and at about the same time. If he had published his results as they stood, he would have given the world conclusions almost identical with those that Lavoisier was then preparing to communicate to the Academy (Easter 1775).

On the 8th of this month I procured a mouse, and put it into a glass vessel, containing two ounce-measures of the air from mercurius calcinatus. Had it been common air, a full-grown mouse, as this was, would have lived in it about a quarter of an hour. In this air, however, my mouse lived a full half hour; and though it was taken out seemingly dead, it appeared to have been only exceedingly chilled; for, upon being held to the fire, it presently revived, and appeared not to have received any harm from the experiment.

By this I was confirmed in my conclusion, that the air extracted from mercurius calcinatus, &c. was, *at least, as good* as common air; but I did not certainly conclude that it was any *better*; because, though one mouse would live only a quarter of an hour in a given quantity of air, I knew it was not impossible but that another mouse might have lived in it half an hour; so little accuracy is there in this method of ascertaining the goodness of air: and indeed I have never had recourse to it for my own satisfaction, since the discovery of that most ready, accurate, and elegant test that nitrous air furnishes. [This is an example of sound and careful reasoning about the tactics of experimentation.] But in this case I had a view to publishing the most generally-satisfactory account of my experiments that the nature of the thing would admit of.

This experiment with the mouse, when I had reflected upon it some time, gave me so much suspicion that the air into which I had put it was better than common air, that I was induced, the day after, to apply the test of nitrous air to a small part of that very quantity of air which the mouse had breathed so long; so that, had it been common air, I was satisfied it must have been very nearly, if not altogether, as noxious as possible, so as not to be affected by nitrous air; when, to my surprize again, I found that though it had been breathed so long, it was still better than common air. [Here we come near to the crucial point, the examination of a residual gas after a test. If common air was at hand, the exhaustion of the oxygen by the respiration of the mouse would have yielded an air whose goodness by the "nitrous air test" would have been low.] For after mixing it with nitrous air, in the usual proportion of two to one, it was diminished in the proportion of 4½ to 3½; that is, the nitrous air had made it two ninths less than before, and this in a very short space of time; whereas I had never found that, in the longest time, any common air was reduced more than one fifth of its bulk by any proportion of nitrous air, nor more than one fourth by any phlogistic process whatever. [The difference between two ninths and one fifth (one forty-fifth) was beyond the limits of accuracy of Priestley's measurements. The greatest change in volume by absorption of all the oxygen by combustion (a phlogistic process) yields a diminution of one fifth only if the experiment is carefully performed.] Thinking of this extraordinary fact upon my pillow, the next morning I put another measure of nitrous air to the same mixture, and, to my utter astonishment, found that it was farther diminished to almost one half of its original quantity. I then put a third measure to it; but this did not diminish it any farther: but, however, left it one measure less than it was even after the mouse had been taken out of it. [Here is the crucial experiment.]

Being now fully satisfied that this air, even after the mouse had breathed it half an hour, was much better than common air; and having a quantity of it still left, sufficient for the experiment, viz. an ounce-measure and a half, I put the mouse into it; when I observed that it seemed to feel no shock upon being put into it, evident signs of which would have been visible, if the air had not been very wholesome; but that it remained perfectly at its ease another full half hour, when I took it out quite lively and vigorous. Measuring the air the next day, I found it to be reduced from 1½ to ⅔ of an ounce-measure. And after this, if I remember well (for in my *register* of the day I only find it noted, that it was *considerably diminished* by nitrous air) it was nearly as good as common air. It was evident, indeed, from the mouse having been taken out quite vigorous, that the air could not have been rendered very noxious.

For my farther satisfaction I procured another mouse, and putting it into less than two ounce-measures of air extracted from mercurius calcinatus and air from red precipitate [red oxide prepared via nitric acid] (which, having found them to be of the same quality, I had mixed together) it lived three quarters of an hour. But not having had the precaution to set the vessel in a warm place, I suspect that the mouse died of cold. However, as it had lived three times as long as it could probably have lived in the same quantity of common air, and I did not expect much accuracy from this kind of test, I did not think it necessary to make any more experiments with mice.

Being now fully satisfied of the superior goodness of this kind of air, I proceeded to measure that degree of purity, with as much accuracy as I could, by the test of nitrous air; and I began with putting one measure of nitrous air to two measures of this air, as if I had been examining common air; and now I observed that the diminution was evidently greater than common air would have suffered by the same treatment. A second measure of nitrous air reduced it to two thirds of its original quantity, and a third measure to one half. Suspecting that the diminution could not proceed much farther, I then added only half a measure of nitrous air, by which it was diminished still more; but not much, and another half measure made it more than half of its original quantity; so that, in this case, two measures of this air took more than two measures of nitrous air, and yet remained less than half of what it was. Five measures brought it pretty exactly to its original dimensions. [It is assumed that the experiments are carried out over water, of course.]

At the same time, air from the *red precipitate* was diminished in the same proportion as that from *mercurius calcinatus*, five measures of nitrous air being received by two measures of this without any increase of dimensions. Now as common air takes about one half of its bulk of nitrous air, before it begins to receive any addition to its dimensions from more nitrous air, and this air took more than four half-measures before it ceased to be diminished by more nitrous air, and even five half-measures made no addition to its original dimensions, I conclude that it was between four and five times as good as common air. It will be seen that I have since procured air better than this, even between five and six times as good as the best common air that I have ever met with.

These two paragraphs record the convincing experiments which showed that the gas at hand was not common air but a new substance. Pure "nitrous air" and pure oxygen when brought together over water combine in about the ratio of 2 to 1; the exact relation depends on the way the experiment is performed and the temperature. Priestley's results are the equivalent of a reaction between about 4 half measures or 2 volumes of nitrous air and 1 volume of oxygen. This indicates that his gases were quite pure. The final residual gas was the excess of nitrous air which was used. If he had used smaller increments he might have had almost no residue after the proper amount of nitrous air had been added over water. The effective discovery of oxygen by Priestley may be considered to have taken place when these experiments were made and interpreted.

Being now fully satisfied with respect to the *nature* of this new species of air, viz. that, being capable of taking more phlogiston from nitrous air, it therefore originally contains less of this principle [a consistent use of the phlogiston terminology]; my next inquiry was, by what means it comes to be so pure, or philosophically speaking, to be so much *dephlogisticated*; and since the red lead yields the same kind of air with mercurius calcinatus, though mixed with common air, and is a much cheaper material, I proceeded to examine all the preparations of lead, made by heat in the open air, to see what kind of air they would yield, beginning with the *grey calx*, and ending with *litharge*.

The red lead which I used for this purpose yielded a considerable quantity of dephlogisticated air, and very little fixed air; but to what circumstance in the preparation of this lead, or in the keeping of it, this difference is owing, I cannot tell. I have frequently found a very remarkable difference between different specimens of red lead in this respect, as well as in the purity of the air which they contain. [Here Priestley practically admits that his samples of red lead were not homogeneous, but neither then nor subsequently does he draw the correct conclusion. This is another illustration of the difficulties of correct interpretation of chemical "facts."] This difference, however, may arise in a great measure, from the care that is taken to extract the fixed air from it. In this experiment two measures of nitrous air being put to one measure of this air, reduced it to one third of what it was at first, and nearly three times its bulk of nitrous air made very little addition to its original dimensions; so that this air was exceedingly pure, and better than any that I had procured before.

Then there follows in Priestley's volume an account of many experiments with the oxides of lead. Of these there are a number containing, as we now know, different ratios of oxygen and lead atoms. They have different colors which serve to distinguish them from each other. As prepared by the various methods known to Priestley, they often contained not only a mixture of lead oxides but also lead carbonate and lead nitrate (if nitric acid was used in their preparation). Probably most, if not all, the oxygen that Priestley obtained from them came from the decomposition of the lead nitrate, for nitrates on heating frequently yield oxygen, an oxide of nitrogen, and an oxide of the metal. We need not pursue Priestley's investigation of these complications further except to note that he was led farther and farther from the right road by these subsequent inquiries. He began to confuse nitrates and carbonates and oxides and consequently developed the idea that a calx had imbibed "spirit of nitre," that is, nitric acid, from the air. *Not having any adequate criteria of purity and not operating with constant regard for changes in weight and in volume, he proceeded from one faulty experiment to another.* We need not follow his further work on calces, for he never arrived at a final conclusion. But he always had at his disposal some alleged fact, such as the statement that "some metallic calces yield fixed air on heating," to challenge Lavoisier's conceptual scheme and support the phlogiston theory. For the student of the history of the development of science, the importance of this part of the story, thus briefly summarized, is as follows: homogeneity of materials is as essential to the chemist as the control of such variables as temperature and pressure is to the physicist; furthermore, criteria for determining homogeneity are often extremely difficult to establish, and it was only slowly that such criteria were evolved. Only after Lavoisier's "principle of the balance sheet" became accepted among chemists could such criteria be found for many materials. A *complete quantitative* study of the decomposition by heat of various samples of red lead, for example, with and without the addition of charcoal, will yield the necessary information to tell the investigator whether two samples are identical in composition and whether carbonate or nitrate is also present.

### 5. NOTES ON PRECEDING SECTIONS

A study of the accounts by Lavoisier and Priestley of their first experiments with oxygen illustrates a number of general principles in the development of science. *First,* a contrast of the method of publishing the results and the reference to other works with early seventeenth-century practice on the one hand and twentieth-century procedure on the other, illustrates the growth of science as an organized activity. *Second,* the difficulties of chemical experimentation are exposed very clearly; the difficulties are sometimes those of interpretation of what is observed, sometimes the failure to try what now seems an obvious further experiment, often the failure to have homogeneous materials at hand. *Third,* the role of the accidental discovery is almost glorified by Priestley. *Fourth,* repeated use of the *limited* working hypothesis is evident. For example, every time a chemical test is applied, Priestley or Lavoisier is essentially saying, "If I do so and so, such and such will happen." Priestley's original faulty identification of oxygen as laughing gas and his failure to interpret the "nitrous air test" correctly shows how many hidden assumptions are involved in the interpretation of experimental results. *Fifth,* Priestley's blind adherence to the phlogiston theory in spite of his own effective discovery of oxygen and in spite of its obvious faults (such as the failure to account for the increase in weight on calcination) shows the hold that one conceptual scheme may have on the mind of an investigator. *Sixth,* the transformation of a broad working hypothesis into a new conceptual scheme is made evident by following Lavoisier's work. His broad working hypothesis was by 1778 well on its way to becoming a new conceptual scheme of revolutionary importance.

Long before Priestley and Lavoisier, the increase in weight of metals on calcination was recorded and at least one investigator had come close to an effective discovery of oxygen. Priestley's emphasis on the accidental nature of scientific discoveries should not deceive the reader into believing that scientific progress hangs on such accidents, nor should Lavoisier's brilliance lead one to conclude that science is solely the work of a few great men. Among the complex of conditions which determined that in the late eighteenth rather than in the late seventeenth century the time was ripe for an elucidation of combustion we may list the following: (*a*) the improvement in communications among scientific men, which made science more and more of a cooperative effort; (*b*) the accumulation of quantitative studies in physics that made unsatisfactory the concept of phlogiston, which implied a substance with a negative weight; (*c*) the accumulation of a century's work on the materials, apparatus, and techniques of chemistry.

### 6. THE COMPOSITION OF THE ATMOSPHERE

The discovery of oxygen was the central event in the overthrow of the phlogiston theory. But it must be remembered that it was the discovery that *oxygen was a constituent of the atmosphere* which provided the key to the riddle of combustion. The method of preparing the red oxide of mercury was an essential link in Lavoisier's argument. He was at some pains in his Easter Memoir to show that the red powder was a true calx; it was formed when mercury was heated in air. In this process the gain in weight was due to combination of either air or a constituent of air with the mercury. By 1777 Lavoisier was clear that the principle that combined with metals during calcination was "an eminently respirable air" which constituted only a part of common air. In his textbook, *Traité Élémentaire de Chimie* (Elements of Chemistry), published in 1789, he describes an experiment which taken together with the preparation of oxygen from the red oxide demonstrates the composition of the atmosphere by analysis. The following extract is from an English translation of this book by Robert Kerr (London, ed. 5, 1802), modernized by the editor of this Case.

### Lavoisier's *Elements of Chemistry,* Chapter III

It appears that our atmosphere is composed of a mixture of substances capable of retaining the gaseous state at common temperatures, and under the usual degrees of pressure. These gases constitute a mass, in some measure homogeneous, extending from the surface of the earth to the greatest height hitherto attained, of which the density continually decreases in the inverse ratio of the superincumbent weight. . . .

It is our business to endeavor to determine, by experiments, the nature of the elastic fluids which compose the lower stratum of air which we inhabit. Modern chemistry has made great advances in this research; and it will appear, by the following details, that the analysis of atmospherical air has been more rigorously determined than that of any other substance of the class.

Chemistry affords two general methods of determining the constituent principles of bodies, the method of analysis, and that of synthesis. When, for instance, by combining water with alcohol, we form the species of liquor called, in commercial language, brandy, or spirit of wine, we certainly have a right to conclude, that brandy, or spirit of wine, is composed of alcohol combined with water. We can procure the same result by the analytical method; and in general it ought to be considered as a principle in chemical science, never to rest satisfied without both these species of proofs. We have this advantage in the analysis of atmospherical air; being able both to decompound it, and to form it anew in the most satisfactory manner. I shall, however, at present confine myself to recount such experiments as are most conclusive upon this head; and I may consider most of these as my own, having either first invented them, or having repeated those of others, intended for analyzing atmospherical air, in perfectly new points of view.

I took a flask of about 36 cubical inches capacity, having a long neck bent as shown in the figure [Fig. 3] and placed in the furnace in such a manner that the extremity of its neck might be inserted under a bell-

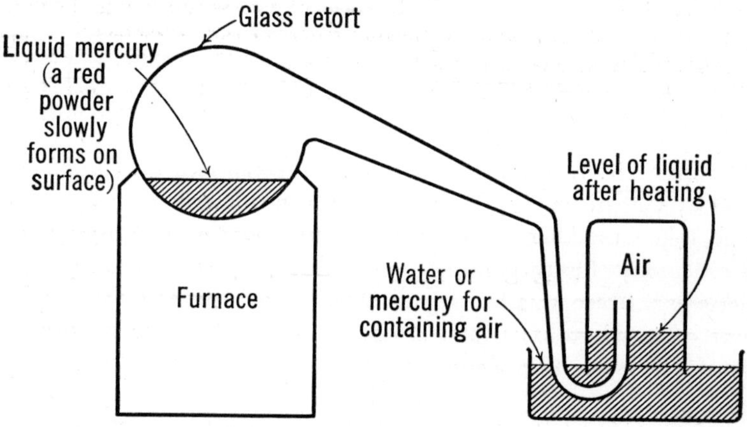


Fig. 3. Boiling mercury absorbs oxygen from the air.

glass, placed in a trough of quicksilver. I introduced four ounces of pure mercury into the flask and by means of a siphon, exhausted the air in the receiver so as to raise the quicksilver, and I carefully marked the height at which it stood, by pasting on a slip of paper. Having accurately noted the height of the thermometer and barometer, I lighted a fire in the furnace which I kept up almost continually during twelve days, so as to keep the quicksilver always very near its boiling point. Nothing remarkable took place during the first day: The mercury, though not boiling, was continually evaporating, and covered the interior surface of the vessel with small drops, which gradually augmenting to a sufficient size, fell back into the mass at the bottom of the vessel. On the second day, small red particles began to appear on the surface of the mercury; these, during the four or five following days, gradually increased in size and number, after which they ceased to increase in either respect. At the end of twelve days, feeling that the calcination of the mercury did not at all increase, I extinguished the fire, and allowed the vessels to cool.

> The bulk of air in the body and neck of the flask and in the bell-glass, reduced to a medium of 28 inches of the barometer and 54.5° of the thermometer, at the commencement of the experiment was about 50 cubical inches. At the end of the experiment the remaining air, reduced to the same medium pressure and temperature, was only between 42 and 43 cubical inches; consequently it had lost about 1/6 of its bulk. Afterwards, having collected all the red particles, formed during the experiment, from the running mercury in which they floated, I found these to amount to 45 grains.
>
> I was obliged to repeat this experiment several times, as it is difficult, in one experiment, both to preserve the whole air upon which we operate, and to collect the whole of the red particles, or calx of mercury, which is formed during the calcination. . . .
>
> The air which remained after the calcination of the mercury in this experiment, and which was reduced to 5/6 of its former bulk, was no longer fit either for respiration or for combustion: animals being introduced into it were suffocated in a few seconds, and when a taper was plunged into it, it was extinguished, as if it had been immersed in water.

This gas was commonly called "mephitic air." Lavoisier named it azote; in English the name nitrogen was introduced at this time.

Lavoisier then describes the famous experiment of the Easter Memoir but uses only the 45 grains of red oxide prepared in the preceding experiment. From this he obtains 7 to 8 cubic inches of gas and 41.5 grains of mercury. It is interesting to compare this result with those reported in 1775. The weight of the gas by difference is 3.5 grains, which gives a weight per cubic inch of about 0.5 or 0.44 grain (3.5 divided by 7 or 8). This figure is to be compared with the nearly 2/3 reported earlier (p. 26). At this point an analysis of the atmospheric air had been accomplished. By the aid of mercury, a furnace, and other equipment, 50 cubic inches of air had been separated into 42 to 43 cubic inches of air unfit for respiration or combustion (nitrogen) and 7 to 8 cubic inches of oxygen.

Lavoisier at this point draws the attention of the reader to the conclusion "that atmospheric air is composed of two elastic fluids (gases) of different and opposite qualities." He further adds, "As a proof of this important truth, if we recombine these two elastic fluids which we have separately obtained in the above experiment, viz. the 42 cubical inches of mephitic air [nitrogen] with the 8 cubical inches of highly respirable air, we reproduce an air precisely similar to that of the atmosphere, and possessing nearly the same power of supporting combustion and respiration, and of contributing to the calcination of metals." This is proof by synthesis as Lavoisier used the word in his book. Since air is a mixture and not a chemical compound, some chemists today would not speak of synthesizing air by mixing nitrogen and oxygen, for we generally reserve the word synthesis for the preparation of a chemical *compound*.

Lavoisier in the same chapter speaks of the "mutual adhesion of the two constituent parts of the atmosphere for each other." He thus seems to have something in the nature of a chemical compound in mind. And elsewhere he says that the method of analysis using boiling mercury is not accurate, for "the attraction of mercury to its respirable part of the air or rather to its base (i.e., the gas deprived of caloric fluid) is not sufficiently strong to overcome all the circumstances which oppose this union . . . In consequence, when the calcination ends, or is at least carried as far as is possible in a determinate quantity of atmospheric air, there still remains a portion of respirable air united to the mephitic air which the mercury cannot separate." These statements indicate that Lavoisier's picture of the atmosphere was somewhat different from what we have today. Yet we must recall that in the third paragraph of his Chapter III (see p. 49) he refers to mixing water and alcohol when speaking of a synthesis. The distinction between a mixture and a chemical compound was not yet quite clear. Only by the assiduous use of the "principle of the balance sheet" by hard-working investigators in the next two decades was it finally shown that elements unite in definite proportions to form a compound. A mixture, on the other hand, is characterized by the fact that a little more or less of one component will not greatly alter the properties, as in the case of mixtures of oxygen and nitrogen. Other criteria, of which Lavoisier was unaware, were later developed for distinguishing between mixtures and compounds.

Lavoisier was quite right about the difficulties of his method of analysis. However, the value he obtained of 16 parts by volume of oxygen and 84 of nitrogen is wrong. Dry air is composed of 21 percent of oxygen by volume, 78 percent of nitrogen, nearly 1 percent of the rare gases, and less than 0.05 percent of carbon dioxide. (The amount of moisture present depends on the temperature and the relative humidity. At room temperature the maximum amount of water by volume that can be present is a little over 2 percent. Therefore, the minimum value for the oxygen content in a sample of common air taken without drying, as Lavoisier did, would be about 20.5 percent by volume.) Accurate methods of determining the composition of air were developed shortly after Lavoisier's first experiments and were improved in the following century. In principle they are identical with his method but a chemical substance is employed (usually in solution) that will rapidly combine with the oxygen present at room temperature. The change in volume, as in Lavoisier's experiment, is a measure of the oxygen present.

### 7. THE FINAL COLLAPSE OF THE PHLOGISTON THEORY

Although Lavoisier's new conceptual scheme was clearly developed and well buttressed by experiments by 1778, he did not attack the prevailing doctrine directly until 1783. In that year the French chemist published his *Reflections on Phlogiston*, which marshaled the evidence for the new ideas and showed that the concept of phlogiston was not only unnecessary but self-contradictory. At just this time the composition of water was established by experiments of Henry Cavendish (1731–1810) which were immediately repeated by Lavoisier. Priestley had once again started a series of important experiments but had not in this case carried through to even a correct qualitative conclusion. Historians of chemistry still argue over who is entitled to the credit for the highly important discovery that water is composed of oxygen and hydrogen in the proportions by weight of approximately 8 to 1. Priestley, Cavendish, Lavoisier, and James Watt of steam-engine fame, are all contenders for the honor.

With the discovery that water was formed when hydrogen was burned in air, Lavoisier's scheme was complete. Water was clearly the oxide of hydrogen. Lavoisier at once proceeded to test an obvious deduction from this extension of his conceptual scheme, namely, that steam heated with a metal should yield a calx and hydrogen. It did. (The converse was likewise demonstrated at about the same time.)

Hydrogen + Oxygen → Water
Steam heated with metal → Calx + Hydrogen
(oxide)

These facts about the relation of water, hydrogen, oxygen, metals, and oxides would seem to leave no ground for the supporters of the phlogiston theory to stand on. But for a few years the new knowledge had the contrary effect. The believers in phlogiston were at last able to explain why a calx weighed more than the metal. This they did by a modification of the phlogiston theory which is illustrated by the following table:

*Modified Phlogiston Theory*
(about 1785)

Hydrogen = phlogiston (often carrying water);
Oxygen = dephlogisticated air;
Water = dephlogisticated air + phlogiston;
Nitrogen = completely phlogisticated air;
Common air = partially phlogisticated air carrying water;
Metal = calx + phlogiston − water;
Calx = the base of a pure earth + water;
Charcoal = phlogiston + ash + water.

If one studies this table it becomes clear that a fairly satisfactory account can be given of the simple chemical reactions. Thus in calcination the following process took place:

Metal heated in air → Phlogisticated air + Calx (water absorbed from the air).

The metal lost phlogiston to the air, which would mean a loss in weight, *but* the resulting "base of the pure earth" absorbed water from the air and the consequent gain in weight more than offset the loss in weight due to the loss of phlogiston!

The classic experiment with red oxide of mercury was formulated somewhat as follows: red oxide is a simple substance containing water; on heating, the phlogiston of the water combines with the simple substance (the pure earth) yielding the metal, while the rest of the water (dephlogisticated air) comes off and is collected. When mercury is heated in oxygen, the reverse process occurs; the phlogiston combines with the dephlogisticated air, forming water, which unites with the simple substance to form the calx.

Henry Cavendish, one of the great scientific figures of the period (though a most eccentric gentleman), was for a time an adherent of the modified phlogiston theory. In a famous article published in 1784 (in the *Philosophical Transactions of the Royal Society,* Vol. 74) he summarizes Lavoisier's new ideas quite fairly and then writes (pp. 151–152):

"It seems, therefore, from what has been said, as if the phenomena of nature might be explained very well on this [i.e., Lavoisier's] principle without the help of phlogiston; and indeed, *as adding dephlogisticated air to a body comes to the same thing as depriving it of its phlogiston and adding water to it,* and as there are, perhaps, no bodies entirely *destitute* of water, and *as I know no way by which phlogiston can be transferred from one body to another, without leav-*

*ing it uncertain whether water is not at the same time transferred,* it will be very difficult to determine by experiment which of these opinions is the truest; but as the commonly received principle of phlogiston explains all phenomena *at least as well* as Mr. Lavoisier's, I have adhered to that" [italics by the editor of this Case].

In retrospect, we can see that the adherents to the modified phlogiston theory were fighting a rear-guard action. Before Lavoisier's execution by the revolutionary tribunal in 1794, many chemists had come to accept his views. By the end of the century Priestley was almost alone in defending the doctrine of phlogiston. The story of the last days of the phlogiston theory is of interest, however, in illustrating a recurring pattern in the history of science. It is often possible by adding a number of new special auxiliary postulates to a conceptual scheme to save the theory — at least temporarily. Sometimes, so modified, the conceptual scheme has a long life and is very fruitful; sometimes, as in the case of the phlogiston theory after 1785, so many new assumptions have to be added year by year that the structure collapses. Most of the illustrations of this pattern, it should be pointed out, concern concepts and conceptual schemes of far less breadth than the phlogiston doctrine. They may be ideas that are useful in formulating merely some relatively narrow segment of physics, chemistry, astronomy, or experimental biology. What has just been said applies none the less.

The publication of Lavoisier's *Traité Élémentaire de Chimie,* with his exposition of the evidence in support of the new views and his new nomenclature, made the destruction of the phlogiston theory inevitable.

### 8. EIGHTEENTH-CENTURY REVOLUTIONS: CHEMICAL, INDUSTRIAL, POLITICAL, AND SOCIAL

The period 1770–1800 was a time of revolutions. The fact that the chemical revolution was contemporary with the American Revolution and preceded the French Revolution by a few years adds interest to the story. Priestley and Lavoisier were both involved in the French Revolution, the latter losing his life at the height of the terror, the former being driven from England because of his sympathies with the French revolutionists. During the same period the iron and steel industry was undergoing a major transformation that was the key to a series of events often called the industrial revolution of the late eighteenth century. The interconnections between these revolutions — scientific, industrial, political, and social — while many, are far from simple. A study of the lives of Priestley, Lavoisier, and Watt gives one a good picture of the status of scientific activity in this period. The important scientists are still for the most part amateurs, as they were in the days of Robert Boyle (mid-seventeenth century). Not until after the Napoleonic wars were the major contributions to the advance of chemistry (or physics) to come from the laboratories of the universities. The importance of the Royal Society and the French Academy as media of communication of scientific information is evident to all who have read the preceding section on the effective discovery of oxygen.

For those who are interested in the relation of science to technology, a study of the period in question will prove profitable. The advance in science and the progress in the practical arts are both rapid; yet the two borrow relatively little from each other in the way of concepts or new factual information. The cross connections are more evident than a century earlier and both activities have increased in importance since the seventeenth century. The application of science to industry, however, still lies in the future. One must not be misled by coincidences in time and place. For example, it is important to remember that the whole revolution of the making of iron and steel in the eighteenth century was based on purely empirical experimentation. The substitution of coke for charcoal, the invention of the crucible steel process, the improvements in iron-ore smelting that yielded pig iron, the development of the puddling process for making malleable iron (mild steel) were all accomplished without benefit of science. Not until after all these technical advances had been made did the impact of Lavoisier's new ideas enable people to recognize that the fundamental *chemical* distinction between iron and steel lay in the carbon content. The cut-and-try methods of practical men were for the time being far more effective in producing practical effects than the work of scientists.

The progress in the manufacture of all manner of products in this period made men alert to the possibilities of employing scientific discoveries. An improved process for manufacturing sulfuric acid yielded a material useful in the textile industry; James Watt recognized (1785) the possibilities of using chlorine gas (then recently discovered) in bleaching textiles. Pharmacists in this period were keenly interested in chemical developments and prepared chemicals for the use not only of doctors but of those interested in experimentation. The story of Cadet's red oxide of mercury (p. 40) is a case in point. One would like to know when yellow phosphorus first became available to a purchaser, for after its discovery in the seventeenth century its manufacture was more or less of a secret and each investigator had to prepare his own material. But Lavoisier in 1772 was able to purchase the theretofore rare substance. (The availability of new materials has often played an important role in the advance of chemistry; radioactive isotopes, which became readily available in the late 1940's, have stimulated a vast amount of work.) Scientists and inventors were in communication in this period and willingly learned from one another. However, the theoretical framework of physics and chemistry was so meager that the advances in science, except for mechanics, made little impact on the practical arts. Another fifty to seventy-five years of scientific work would be required before the applications of science were of prime importance.

---

The following chronological table may be of interest as a method of relating the various late-eighteenth-century revolutions to one another.

1760 George II dies; accession of George III.

1760 Smeaton improves blast furnaces for producing cast iron.

1760 Bridgewater canal completed; halves cost of coal in Manchester.

1760 Factory system and use of water-driven machinery firmly established in English silk industry.

1764 Hargreaves' spinning jenny introduced into English textile industry.

1765 Stamp Act.

1766 Cavendish (London) isolates and describes hydrogen gas (inflammable air).

1769 Watt's first steam engine. Arkwright greatly improves water-powered machines for textile industry.

1772 Lavoisier's experiments with sulfur and phosphorus. Priestley publishes his nitrous air test for the "goodness" of air.

1774 Death of Louis XV.

1775 Lavoisier's "Easter Memoir" on calcination. Priestley's effective discovery of oxygen.

1776 Declaration of American Independence.

1776 Publication of Adam Smith's *Wealth of Nations.*

1778 Lavoisier's revised memoir on calcination.

1779 Further improvements in textile machinery in England.

1781 Capitulation of British at Yorktown.

1782–1783 Composition of water established; spread of Lavoisier's ideas; last stand of phlogiston theory.

1783 Lavoisier and Laplace determine large number of specific heats.

1784 Cort's puddling process for making malleable iron from cast iron using coal as fuel.

1783 Peace between Great Britain and the United States.

1784 Watt's improved steam engine.

1786 Watt brings to England from France news of Berthollet's process for bleaching by the action of chlorine (then believed to be a compound); this knowledge utilized by English textile industry.

1787 Constitutional convention at Philadelphia.

1789 Publication of Lavoisier's *Traité Élémentaire de Chimie* setting forth results of the chemical revolution in clear and systematic form.

1789 Louis XVI summons the States General (January); third estate adopts title of National Assembly; fall of the Bastille (July).

1791 Birmingham mob burns Priestley's house.

1792 Attack on the Tuileries, Louis XVI a prisoner; French Republic proclaimed.

1793 Louis XVI executed; Committee of Public Safety in Paris; "Terror" in France begins.

1793 War between Great Britain and France, continues with only short truces until 1815.

1794 Execution of Lavoisier (May 8); fall and death of Robespierre (July); end of Terror.

1796 Production of cast iron in Great Britain 125,000 tons, double the figure of a decade earlier.

1798 Battle of the Nile. Nelson's victory assures supremacy of British fleet.

1798 Publication of *Essay on Population* by Malthus.

1799 Napoleon becomes First Consul.

## SUGGESTED READING

### 1. Books Directly Related to the Effective Discovery of Oxygen

*Torch and Crucible, The Life and Death of Antoine Lavoisier*, by Sidney J. French (Princeton University Press, 1941). A popular account of Lavoisier's work; combination of a biography and exposition of the chemical revolution. Recommended for students of this Case.

*Three Philosophers* (Lavoisier, Priestley, and Cavendish), by W. R. Aykroyd (London, 1935; out of print).

*Antoine Lavoisier*, by Douglas McKie (Philadelphia, 1935; out of print).

*The Eighteenth Century Revolution in Science: The First Phase*, by A. U. Meldrum (Calcutta: Longmans, Green, 1930). This pamphlet, which is difficult to obtain, is an authoritative scholarly review of the work of Lavoisier and Priestley.

*Elements of Chemistry*, by Antoine Lavoisier (English translation by Robert Kerr, 1793; many editions). The first four or five chapters can be read with profit by students of this Case.

*Historical Introduction to Chemistry*, by T. M. Lowry (London: Macmillan, 1936). A good guide to those interested in an elementary presentation of any phase of the history of chemistry since 1700.

*Lectures on Combustion*, by Joseph Priestley and John MacLean, edited by William Foster (Princeton University Press, 1929). Throws interesting light on Priestley's last stand in defense of the phlogiston theory.

### 2. The Eighteenth-Century Background

Among the vast number of books dealing with the political and social history of the last half of the eighteenth century, the following may have special relevance for the student of this Case.

*Life of Joseph Priestley*, by Anne Holt (Oxford, 1931).

*The Eighteenth Century Background*, by Basil Willey (London, 1940). Chapter X on Priestley is recommended.

*Science and Social Welfare in the Age of Newton*, by G. N. Clark (Oxford: Clarendon Press, 1937). Although this little volume deals with an earlier period, it is recommended for students of this Case as providing a picture of the economic and cultural factors involved in the interaction of science and technology.

*Iron and Steel in the Industrial Revolution*, by T. S. Ashton (1924). Too detailed and technical for the average student, but contains a great deal of interesting material.